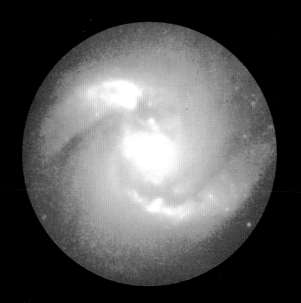

The Star Guide

Second edition

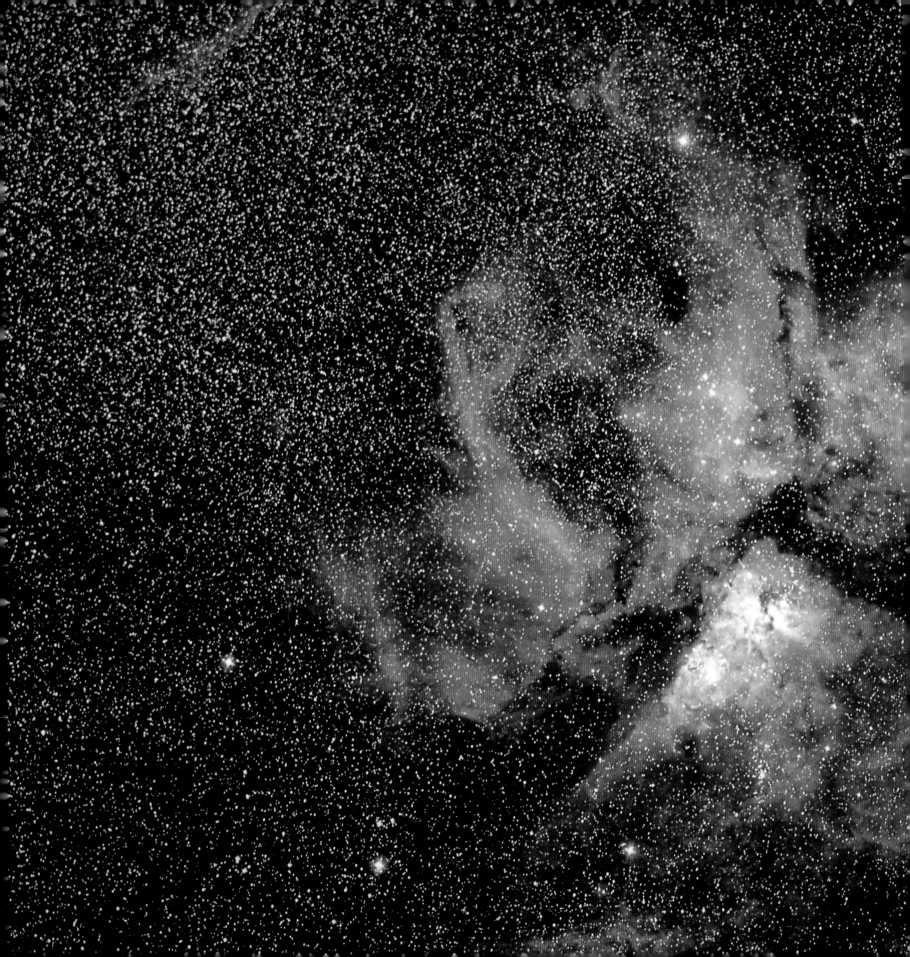

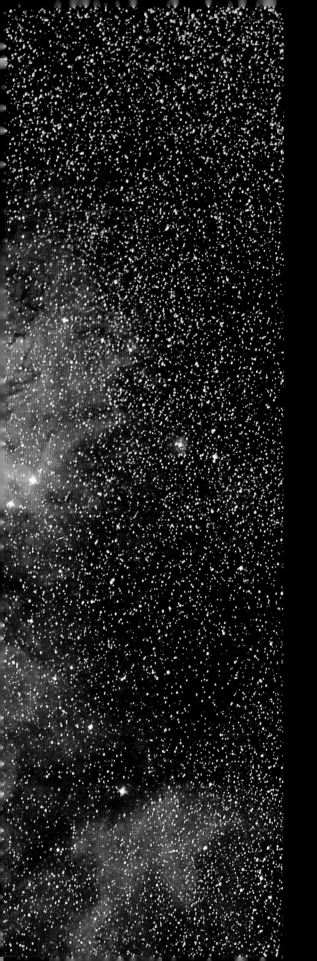
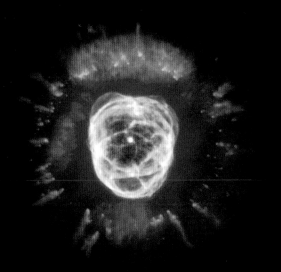

The Star Guide

Second edition

Learn how to read the night sky star by star

Robin Kerrod

A QUARTO BOOK
Copyright © 2005 Quarto Inc.
Published by John Wiley & Sons Inc., Hoboken, New Jersey
No part of this publication may be reproduced, stored in a retrieval system, or transmitted in any form or by any means, electronic, mechanical, photocopying, recording, scanning, or otherwise, except as permitted under Section 107 or 108 of the 1976 United States Copyright act, without either the prior written permission of the publisher, or authorization through payment of the per-copy fee to the Copyright Clearance Center Inc., 222 Rosewood Drive, Danvers, MA 01923, (978) 750-8400, fax (978) 750-4470, or on the web at www.copyright.com.

Requests to the Publisher for permission should be addressed to Permissions Department, John Wiley & Sons, Inc., 111 River Street, Hoboken, NJ 07030, (201) 748-6011, fax (201) 748-6008, e-mail: permcoodinator@wiley.com.

Limit of Liability/Disclaimer of Warranty: While the publisher and author have used their best efforts in preparing this book, they make no representations or warranties with respect to the accuracy or completeness of the contents of this book and specifically disclaim any implied warranties of merchantability or fitness for a particular purpose. No warranty may be created or extended by sales representatives or written sales material. The advice and strategies contained herein may not be suitable for your situation. You should consult with a professional where appropriate. Neither the publisher nor author shall be liable for any loss or profit or any other commercial damages, including but not limited to special, incidental, consequential, or other damages.

Library of Congress Cataloging-in-Publication Data
Kerrod, Robin.
  The star guide : learn how to read the night sky star by star / Robin Kerrod.-- 2nd ed.
      p. cm.
  Includes index.
  ISBN 0-471-70617-5 (cloth)
  1. Stars--Observers' manuals. 2. Astronomy--Amateurs' manuals. I. Title.
  QB63.K45 2005
  523.8'022'3--dc22
2004022953

Conceived, designed, and produced by
Quarto Publishing plc
The Old Brewery
6 Blundell Street
London N7 9BH

Art Editor: James Lawrence
Assistant Art Director: Penny Cobb
Project Editor: Trisha Telep
Illustrator: Kuo Kang Chen

Publisher Piers Spence
Art Director Moira Clinch

Manufactured by Modern Age Repro House Ltd, Hong Kong
Printed by SNP Leefung printer Limited in China

Contents

Introduction	6
How To Use This Book	8
1 Observing the Heavens	**10**
Stargazing	12
Realm of the Big Eyes	14
Photographing the Heavens	16
Radio and Space Telescopes	18
The Hubble Space Telescope	20
2 Looking at Stars	**22**
Distant Suns	24
The Message in Starlight	26
Clouds Among the Stars	28
Star Birth	30
Star Death	32
Star Groups and Clusters	34
The Milky Way	36
Our Galactic Neighbors	38
Galaxies Galore	40
Hyperactive Galaxies	42
Galaxies and the Universe	44
3 Patterns in the Sky	**46**
The Constellations	48
Signposts to the Stars	50
The Celestial Sphere	52
Changing Skies	54
Northern Hemisphere, Winter	56

Northern Hemisphere, Summer	58
Southern Hemisphere, Winter	60
Southern Hemisphere, Summer	62

4 The Skies Month By Month — 64

The Monthly Maps	66
North Polar Stars	68
Cassiopeia	70
South Polar Stars	72
Centaurus/Crux	74
January Skies	76
Orion	78
February Skies	80
March Skies	82
Ursa Major	84
April Skies	86
May Skies	88
Virgo	90
June Skies	92
Scorpius	94
July Skies	96
Sagittarius	98
August Skies	100
Cygnus	102
September Skies	104
October Skies	106
Andromeda	108
November Skies	110
December Skies	112
Taurus	114

5 Sun and Moon — 116

Daytime Star	118
Observing Solar Eclipses	120
Queen of the Night	122
Moon Movements	124
Lunar Landscapes	126
Northwest Quadrant	128
Southwest Quadrant	130
Northeast Quadrant	132
Southeast Quadrant	134

6 The Sun's Family — 136

The Solar System	138
Observing the Planets	140
The Scorching Planets	142
The Red Planet	144
The Giant Planets	146
Far Distant Worlds	148
Many Moons	150
Meteors and Comets	152
The Asteroids	154
Glossary	156
Index	158
Credits	160

LEFT **KITT PEAK NATIONAL OBSERVATORY**
Located near Tucson, in Arizona.

Introduction

Long before the dawn of civilization, our early ancestors would have gazed with awe and wonderment at the starry heavens that created the great celestial dome above them every night. And even today, when we think we know what makes the heavens—and the vast hidden Universe far beyond the visible stars—tick, we can still gaze in total fascination at the night sky.

We can follow the march of the constellations across the sky, night by night and season by season during the year. We can see meteors flash by nearly every night, looking like stars plummeting to Earth from outer space. Far less often, but much more spectacularly, the maverick bodies we call comets can grace our skies for months, with long tails that can stretch halfway across the heavens. Much more spectacular still are the visual delights that occur when the Moon covers up the Sun to create solar eclipses, or Earth's shadow passes over the Moon to create lunar eclipses. To view all these astronomical phenomena, we need only our eyes, binoculars, and the simplest of telescopes.

Simple stargazing provides the main thrust of the book. However, we also peer beyond the constellations and look in some detail at what the Universe is really like and how it works. In the introductory information pages, we review how astronomers gather their information using giant telescopes on the ground and in space. The data and beautiful images they acquire gives astronomers an ever-increasing understanding of our enigmatic and boundless Universe.

So enjoy the night sky and its myriad stars. Let your mind ponder the curious bodies you might encounter there, from stars suffering birth pangs or dying in supernova explosions, to enigmatic bodies like quasars and blazars.

And, whenever you go stargazing, may you have clear skies.

How to use this book

The Star Guide *has been designed to be as user-friendly as possible. Would-be observers can start stargazing immediately using the monthly maps. But extra information about the kinds of sky objects you see is always available, using the novel icon-based cross-referencing system. Chapter 1 discusses how astronomers gather information, while Chapter 2 looks at the results of their investigations. Chapter 3 contains the first of the star maps. Chapter 4, the main mapping section, tracks the constellations month by month.*

In the main, the book is organized on a spread-by-spread basis, with each two-page spread covering a particular aspect of the night sky or presenting reference information on a specific topic. The first two chapters provide an overview of modern astronomy, looking at the work of astronomers, and providing an introduction to the Universe as we know it today.

BELOW **SEASONAL STAR MAPS**
The whole-sky maps in Chapter 3 present the first level of information about the night sky. They help observers familiarize themselves with the constellations that come into view season by season. Half-circle sketch maps are included to help interpretation, showing examples of sky views in a particular direction.

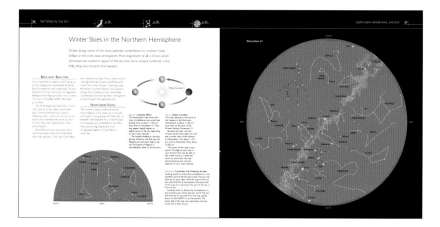

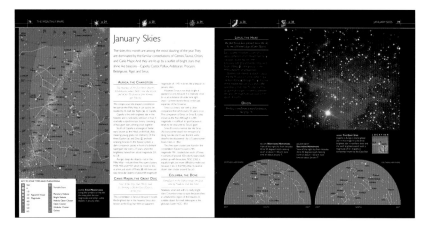

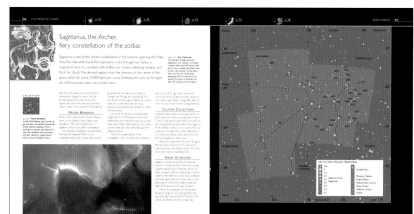

ABOVE **MONTHLY STAR MAPS**
The main maps in Chapter 4 provide more detailed information about the stars and constellations visible month by month throughout the year. They feature the brightest nebulae, clusters, and galaxies. Half-circle sketch maps are again included to help interpretation. They show the kinds of sky views northern and southern observers would see some time during the month.

ABOVE **KEY CONSTELLATIONS**
The key constellation maps in Chapter 4 provide a third level of information about particular constellations, which by their richness merit more detailed investigation than usual by eye, binoculars, and telescope. They include the magnificent Sagittarius, which features dense star clouds, nebulae, and clusters.

THE MAPS

Chapters 3 and 4 feature progressively more detailed star maps. The idea is that observers first become familiar with the shapes and locations of the constellations and then go on to investigate them in greater detail.

The maps in Chapter 3 show how the heavens change in the different seasons—summer and winter. Maps are included for both the Northern and the Southern Hemispheres. Half-circle sky views show the constellations visible looking west.

The main maps in Chapter 4 feature the stars and constellations prominent in the sky in particular months. They also include the brightest deep-sky objects, such as star clusters, nebulae, and galaxies. Half-circle sky views help northern and southern observers orientate themselves. The more detailed maps in Chapter 4 present further information on particularly outstanding constellations. The maps in Chapter 5 feature the Moon. They cover the near side—the one that always faces us—in four quadrants. The seas, the main craters, mountain ranges, and other features of the lunar surface are marked.

CROSS-REFERENCING

The mapping pages are concerned primarily with observation and practical astronomy in general. They can be read and appreciated quite independently of the information pages that appear throughout the book. But if you need further information, you can turn to the relevant pages, using the icons.

If, for example, you are reading about the Orion Nebula and want to know more about nebulae, look for the "Nebula" icon in the band and note the number beside it. Turn to that page and you will find detailed information on nebulae. If, at any time, you want to refresh your memory on common astronomical terms, turn to the Glossary on pages 156 and 157.

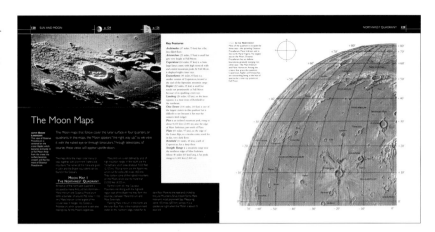

TOP MOON MAPS
The maps in Chapter 5 feature our nearest neighbor in space, the Moon. There are four maps, each covering one quadrant. The maps are printed with north at the top, and show the Moon as we see it with the naked eye and through binoculars. The maria (seas) and most prominent craters are marked.

ABOVE INFORMATION PAGES
This is an example of two information pages. These appear throughout the book and provide in-depth background information on a variety of important topics.

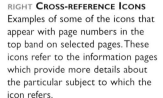

RIGHT CROSS-REFERENCE ICONS
Examples of some of the icons that appear with page numbers in the top band on selected pages. These icons refer to the information pages which provide more details about the particular subject to which the icon refers.

The Milky Way

Observing Solar Eclipses

Galaxies Galore

The Constellations

The Solar System

The Giant Planets

Meteors and Comets

Many Moons

The Hubble Space Telescope

 Observing the heavens

Every clear moonless night, away from bright city lights, nature treats us to a dazzling spectacle. Myriad twinkling stars shine down from the sky, sparkling like jewels against the velvety backdrop of space.

Even in our modern age, we find the night sky as breathtakingly beautiful as our early ancestors did. When they began observing the heavens regularly and noting the seasonal changes among the constellations, they laid the foundations for what is now the meticulous science of astronomy. Exactly when this happened is not known. All we know is that when the first written records appeared, about 5,000 years ago in the Middle East, astronomy was well advanced. The Chaldeans, the Babylonians, and the Egyptians all had priest-astronomers who were skilled observers—with the naked-eye, of course.

It was not until 1609 that the era of modern observational astronomy began, when the Italian scientist Galileo built a telescope and trained it on the heavens for the first time. From that time, broadly speaking, the progress of astronomy has hinged upon the increasing size and ability of telescopes to reveal the Universe in ever more intricate detail. Today, powerful optical and radio telescopes on the ground, astronomy satellites in orbit, and space probes journeying deep into the Solar System are showing us a Universe of enormous complexity—and of extraordinary beauty.

LEFT DAZZLING VISTAS
An in-depth view of a star-forming region in our nearest galactic neighbor in space, the Large Magellanic Cloud, provided by one of the finest astronomical instruments ever built, the Hubble Space Telescope. The cluster of brilliant stars at bottom right is known as Hodge 301. It includes brilliant supergiant stars on the brink of exploding as supernovae.

Stargazing

We all begin stargazing "with the naked-eye," or using our eyes alone, which helps us become familiar with the constellations and the night sky as a whole. But we can find out much more about the heavens with a little help—from binoculars and telescopes.

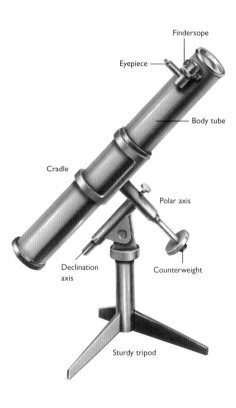

ABOVE **TELESCOPE ANATOMY**
The essential features of the Newtonian reflector are shown in the photograph. The low-powered finderscope helps to zero in on the particular part of the heavens to be studied. The instrument has an equatorial mounting.

LEFT **REFLECTOR**
Explaining the principles of the telescope to a budding astronomer. This model is a Newtonian reflector with an 8.5-inch (21-cm) light-gathering primary mirror. Named after a telescope designed by Isaac Newton in about 1668, it is a good size for the serious observer.

WITH THE NAKED EYE

By simply stargazing using just your eyes, you can see a great deal in the night sky. You can, of course, see masses of stars, the bright ones forming the patterns we call the constellations. The constellations help us find our way around the sky. Altogether, you can see about 2,500 stars above the horizon at any time, although many millions of other stars are also present but too faint for the eye to detect.

What else can you see with the naked eye? Most obviously, the Moon. Earth's constant companion in space, which changes its appearance, goes through its phases over the course of every month. Then there are the bright "wandering stars" that are the planets, with Venus being prominent as an "evening star" on many nights.

Every night you go stargazing, you should see plenty of meteors—fiery trails made by specks of matter from outer space burning up in the atmosphere. Occasionally, you might see one of the most spectacular night-sky objects of all, a comet.

WITH TELESCOPES

The eye is not an ideal instrument for viewing the heavens because it can gather only a small amount of light. Telescopes have a greatly increased capacity for gathering light, making the stars brighter and clearer.

Some telescopes, like the eye, use lenses to gather and focus starlight. Others use mirrors. Telescopes that use lenses are called refractors; those that use mirrors, reflectors. Both produce upside-down images, but this doesn't matter in astronomy.

Refractors have two sets of lenses. An objective lens in front gathers and focuses incoming starlight and forms an image. This image is then viewed and magnified by an eyepiece lens.

Reflectors use sets of mirrors to gather and focus light. In a Newtonian reflector, popular with amateur astronomers, a curved primary mirror gathers light and reflects it to a secondary flat mirror, which in turn reflects it into an eyepiece. The eyepiece is located near the top of the telescope tube.

ABOVE **REFRACTOR**
This refractor has a 4-inch (10-cm) diameter objective lens to gather incoming starlight. At the viewing end, the eyepiece is carried by a right-angled prism mounted in the eyepiece tube, which makes for more comfortable viewing when studying stars high in the sky.

BELOW **AMERICA'S STONEHENGE**
Our early ancestors had only their eyes to view the heavens. But many were expert observers who became familiar with the constellations and the movements of the Moon and Sun. Around 3,500 years ago, Stone Age dwellers living in what is now New Hampshire (near Salisbury) built stone circles with alignments that marked the positions of the rising and setting of the Sun through the seasons. The remains are known as America's Stonehenge, for its resemblance to England's famous stone circle, near Salisbury in southern England.

SIZE MATTERS

The smallest useful size for an amateur refractor is about a 3-inch (75 mm) aperture; for a reflector, about double this size. Refractors are in general more robust and easier to set up than reflectors, but are more expensive.

Binoculars are another invaluable piece of stargazing equipment. Essentially a kind of double refractor, they give a broader field of view than a telescope. They are useful for viewing the Moon, comets, and the nebulae, clusters, and star fields of the Milky Way. They are described by their power of magnification and aperture (in millimeters). Useful sizes for general astronomical observations are 8 X 40, 7 X 50, and 10 X 50. Larger sizes are available but need to be sturdily mounted to prevent image-shake.

ABOVE **GETTING YOUR BEARINGS**
The planisphere is a handy device that helps you get your celestial bearings. (There is one in the pocket at the back of this book.) When you match the time and date of viewing on the scales, the stars and constellations in the sky above are revealed in the "window." Different planispheres are needed for different locations. Those shown here are for latitudes 42 degrees north in the Northern Hemisphere and 35 degrees south in the Southern Hemisphere.

Realm of the Big Eyes

Professional astronomers use reflecting telescopes with mirrors as wide as 33 feet (10 m) to peer deep into the Universe. These telescopes are located at observatories high up on mountaintops, usually in dry climates where they are above the thickest, dirtiest, and most moist layers of atmosphere.

HALE THE GIANT

All the biggest telescopes are reflectors for a very good reason. This is because mirrors can be supported from behind, which helps prevent distortion in large sizes. Glass lenses, however, can be supported only at the edge, so in large sizes they distort under gravity and become optically useless.

The era of giant telescopes began with the completion in 1948 of the Hale Telescope at Mount Palomar Observatory, near Los Angeles. Still in use, its light-gathering primary mirror is 200 inches (5 m) across. Unfortunately, it is suffering increasingly from the blight that is affecting astronomers everywhere—light pollution from urban areas.

ADAPTIVE OPTICS

In modern times, the size and power of the Hale Telescope has been surpassed by many other telescopes. They include the twin Keck reflectors at Mauna Kea Observatory in Hawaii, which have light-gathering mirrors 33 feet (10 m) across. But their mirrors are not built in one piece, because a mirror that size would tend to flex and distort. Instead, the mirror is made up of 36 segments that are individually supported by computer-controlled actuators (rams). The computer continually adjusts the position of each segment so that together they always form the perfect curved shape. This so-called active-optics system compensates for any distortions caused by temperature changes, mirror weight, and even wind load.

Also located at Mauna Kea Observatory are two other large instruments, the 27-foot (8.3-m) Japanese Sabaru Telescope and the slightly smaller Gemini North Telescope.

SOUTHERN GIANTS

Gemini North has an identical twin telescope (hence the name) in Chile, on the mountain Cerro Pachon in the Andes. Like the Kecks, they both have active optics.

The mountain peak of Cerro Paranal in the Atacama Desert in Chile is the home of the European Southern Observatory's Very Large Telescope. This is made up of four identical 27-foot (8.2-m) reflectors. Each instrument can be used to observe on its own, but all four can be electronically linked to act in unison. When all four are used in

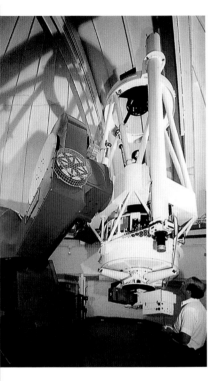

ABOVE IN THE CANARIES Inside the dome of one of the telescopes at the Roque de los Muchachos Observatory on La Palma in the Canary Islands. This compact, easily manipulated modern instrument, known as the Jacobus Kapteyn Reflector, has a mirror 39 inches (1 m) across.

RIGHT PEAK PERFORMANCE Telescope domes at the Kitt Peak National Observatory, near Tucson in Arizona. One of the world's finest observatories, it is located in the Quinlan Mountains above the Sonora Desert. The local Papago Indians call Kitt Peak astronomers "the people with the long eyes."

REALM OF THE BIG EYES

ABOVE **THE VLT**
The four telescopes that make up the Very Large Telescope at Paranal Observatory in Chile. They are named for sky objects in the native Chilean language. Yepun is in the foreground. Behind, from the left, are Antu, Kueyen, and Melipal.

conjunction with other 6-foot (1.8-m) telescopes at the same site, they become equivalent to a reflector with a light-gathering mirror more than 330 feet (100 m) across. It is said that they could spot an astronaut walking on the Moon!

COLOSSAL CAMERAS

While amateur astronomers actually look through their telescopes, professional astronomers seldom do. Instead, they use their instruments as giant cameras to record images on photographic film. Film stores the light falling on it and is thus able to detect very faint stars and galaxies.

Increasingly, however, astronomers use CCDs (charged couple devices) instead of film to record images. CCDs are silicon-chip devices that are much more sensitive to light than ordinary film. Digital cameras and camcorders use similar technology.

Photographing the Heavens

Even with a simple camera, you can emulate professional astronomers and capture images of the night sky on film. Astrophotography—photographing the heavens—can soon become addictive, but don't expect to get the superb color images you see in astronomy books and magazines first time.

Basics

Volumes have been written on the practise of astrophotography, and you are advised to read one if you want to take it up seriously. But here are a few basics.

The first essential is a camera with a facility for time exposure, which is often named a B setting. The reason for this is that the light from the stars is feeble, and the camera shutter must be left open for a while so that the film can "store" the light that falls on it. You may not need a time exposure for the Moon, however, because it is so bright.

You will also need a cable release to activate the camera shutter. This prevents you wobbling the camera when you press the shutter release to expose the film.

The other essential is a sturdy tripod on which to mount your camera. This must hold the camera absolutely steady during a lengthy exposure—if the camera wobbles, the resulting picture will be blurred. Hanging extra weights on a tripod can give it some extra stability.

LEFT HALF MOON
The Moon is a good starting point for budding astrophotographers. Good results can be obtained using a small telescope or even a tripod-mounted camera with a telephoto lens. The picture shows the Moon seven days old, at First Quarter phase. The dark mare (sea) areas contrast markedly with the bright highlands.

ABOVE GOING IN CIRCLES
Star trails in the northern sky are shown here arcing around the Pole Star, Polaris. This star scarcely moves at all because it is less than one degree away from the north celestial pole. This long-exposure photograph was taken on ordinary color film and with an ordinary camera, mounted on a tripod.

LEFT **HALE-BOPP**
If they are bright enough, comets are also easy to photograph with a tripod-mounted camera. The author photographed Comet Hale-Bopp from his garden in March 1997, with an exposure time of about a minute. This was a short enough exposure time not to cause the stars to trail.

FILM FACTS

A wide selection of photographic films is available, with different "speeds." A slow film is less sensitive to light and requires a longer exposure than a fast film. The speed of a film is marked by its ISO (or ASA) rating. You need to select the film speed according to what you are going to photograph.

The simplest pictures you can take are star trails. These are arcs in the sky made by the stars wheeling around the heavens. To record them, point your camera at the sky and open the shutter for a while. For this exercise you can use relatively slow film, such as ISO 100 or 200.

STAR POINTS

If you want to record stars as points of light rather than as trails, you need an exposure of less than a minute. So you will require a faster film, such as ISO 400. This film can also be "pushed," which means that it can be processed as though it were an even faster film, say ISO 800. Much faster films are also available, with ISO ratings of 1,600 and even 3,200. But these tend to produce more grainy images that lack the sharp detail of the slower films.

To improve your stellar photography still further, you need to mount and drive your camera to follow the rotation of the heavens. This will enable you to take lengthy time exposures that will begin to record the splendor of deep-sky objects such as nebulae, star clusters, and galaxies.

If you have a small telescope with a motor drive, you can mount the camera on that. Or you can fit the camera body to the telescope instead of the eyepiece and use the telescope as a supertelephoto lens.

ABOVE **SHOOTING ECLIPSES**
Ordinary cameras can be used to "shoot" eclipses, providing suitable precautions are taken. The author (appropriately dressed!) demonstrates the simple set-up he used for the 1991 total eclipse in Hawaii. For the partial phases, the camera, mounted securely on a tripod, is fitted with an aluminized filter to reduce the Sun's glare. During totality, the filter can be removed to photograph the eclipsed Sun. But it must be put back as soon as daylight returns after the "diamond-ring" stage.

Radio and Space Telescopes

Some of the most spectacular discoveries in astronomy in recent years—such as pulsars and quasars—have been made by telescopes that can pick up the invisible radiations that stars and galaxies give out. Some of these radiations can be picked up from the ground, but most can be detected only from above the atmosphere, in space.

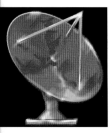

ABOVE **X-Ray Vision**
The satellite ROSAT (Roentgen satellite) detected powerful X-radiation coming from the heart of a cluster of galaxies known as Abell 2256. X-ray emission attends regions of high temperatures and hectic activity in supernova remnants, pulsars, and black holes.

BELOW **Southern Power**
The Parkes radio telescope in New South Wales, Australia. One of the most powerful instruments in the Southern Hemisphere, it has a dish 210 feet (64 m) across.

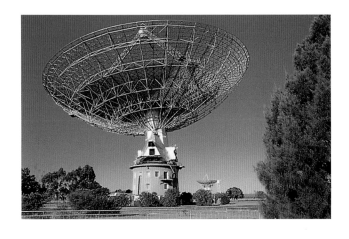

Electromagnetic Waves

Stars and galaxies give out a prodigious amount of energy, not only as light rays we can see, but also as rays, or radiation, we can't see. These invisible radiations include gamma rays, X-rays, ultraviolet rays, infrared rays, microwaves, and radio waves. All these rays and light take the form of electric and magnetic vibrations, which are named electromagnetic waves.

The difference between the various types of radiation lies in their wavelength. Gamma rays have the shortest wavelengths, radio waves the longest.

The stars do not give out energy equally at all wavelengths. For example, a star that looks dim in ordinary visible light may shine like a beacon, say, at radio wavelengths. So, to get a true picture of the heavens, astronomers should ideally view them at all wavelengths. The trouble is that Earth's atmosphere blocks most of the invisible wavelengths out.

BELOW **The Big Dish**
The Arecibo radio telescope in Puerto Rico has the biggest dish in the world. Measuring 1,000 feet (308 m) across, the dish nestles in a natural bowl in a mountaintop. It is made up of 38,778 perforated aluminum panels. It focuses the radio waves it gathers from the heavens onto the antenna suspended high above.

The Radio Window

Fortunately, the atmosphere does let through radio waves, which has led to one of the most exciting branches of astronomy—radio astronomy. This science has its origins in 1931, when Bell Telephone Laboratories researcher Karl Jansky discovered radio waves coming from the heavens.

The most powerful radio telescope is the Very Large Array, near Socorro in New Mexico. It comprises 27 separate dishes mounted on rail tracks, which can be arranged in a number of ways. By linking them together electronically, they can be made to act like a single dish nearly 20 miles (30 km) across.

In a radio telescope, the dish acts as a collector of radio signals and focuses them onto an antenna above it. The signals are then fed through detector circuits like those in a conventional radio set. A computer then processes the signals and can display them as radio "pictures." These pictures are what we might see if our eyes were sensitive to radio waves, and are shown in false (artificially-selected) color.

RADIO AND SPACE TELESCOPES

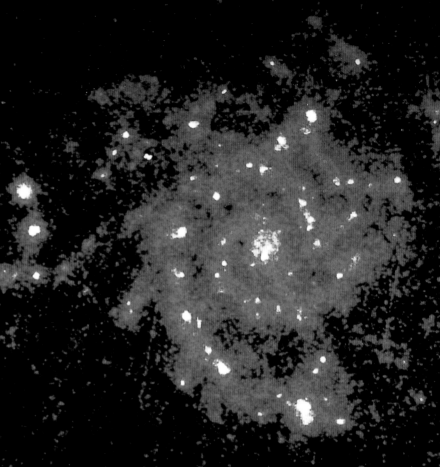

Satellite Observatories

To detect the wavelengths that the atmosphere blocks—gamma rays, X-rays, and so on—astronomers send instruments into space on satellites. (To investigate objects in the Solar System, they send probes, see page 142.)

Among the most notable astronomy satellites have been the Compton Observatory (launched in 1992) and Integral (2002), which studied gamma rays; ROSAT (1990), Chandra (1999), and XMM-Newton (1999), which studied X-rays; IRAS (1983) and ISO (1995), which studied infrared rays; and COBE (1989), which studied microwaves.

But the most outstanding astronomy observatory by far has been the Hubble Space Telescope (1990), which operates primarily at visible-light wavelengths (see page 20).

ABOVE IRAS IN ORBIT
IRAS, the infrared astronomy satellite, pioneered infrared study of the Universe. To make it more sensitive to the feeble radiation from distant objects, its detector was cooled to a temperature just above absolute zero, -459°F (-273°C), by liquid helium.

TOP RIGHT SPARKLING SPIRAL
A shuttle-borne camera took this ultraviolet image of a face-on spiral galaxy known as M74 in the constellation Pisces. The bright regions pinpoint the location of hot newborn stars, which give out intense ultraviolet radiation.

RIGHT ON THE RADIO
A radio image of a distant galaxy. The radio signals come not just from the galaxy itself, but also from regions millions of light-years on either side. This radio galaxy pumps out extraordinary energy at radio wavelengths, which is probably generated by a supermassive black hole at its center.

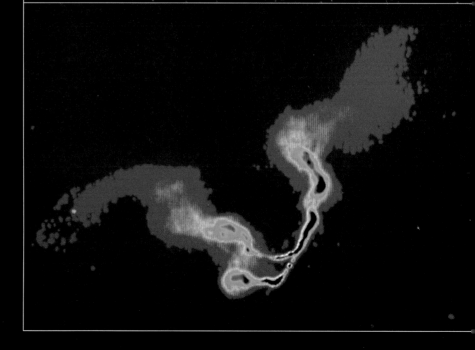

The Hubble Space Telescope

Circling around Earth in orbit, about 350 miles (560 km) high, a satellite as big as a bus has been sending back superlative images of our Universe since its launch in 1990. It is the Hubble Space Telescope, named for U.S. astronomer Edwin Hubble, who pioneered study of the galaxies early last century.

Flawed Vision

After the Hubble Space Telescope (HST) had been launched, on April 24, 1990, mission scientists and astronomers everywhere waited for "first light." This was the time when the HST would send back its first image.

Imagine their dismay when they found that the image was blurred! It transpired that during manufacturing, the telescope's main mirror had been ground to slightly the wrong curvature, and this caused the blurring. Compared with the images from ground telescopes, HST images were good, but nowhere near as good as they should have been.

After much soul-searching, NASA launched a make-or-break servicing mission in December 1993 to correct the HST's flawed vision. On the mission, shuttle astronauts spent more than 35 hours spacewalking to install new equipment. The most vital piece, called COSTAR, used an ingenious arrangement of ten small mirrors to bring light from the misshapen primary mirror into proper focus.

Within days, the HST was sending back some of the best astronomical images ever, at last fulfilling its promise of "opening up a new window on the Universe."

LEFT SERVICING HUBBLE
The Hubble Space Telescope, docked in the payload bay of the shuttle orbiter Discovery during a routine servicing mission in February 1997. The telescope was designed on a modular basis so that equipment modules could be more easily refurbished or replaced.

THE HUBBLE SPACE TELESCOPE

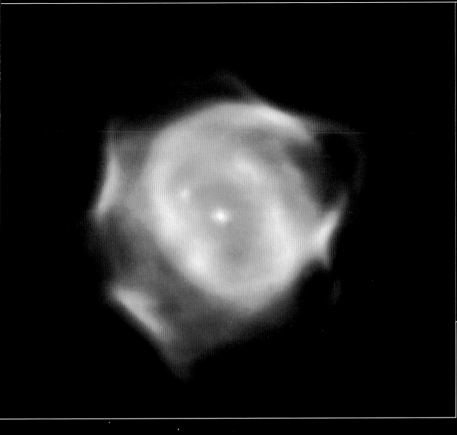

ABOVE THE STINGRAY
Some of the most spectacular images taken by the Hubble Telescope have been of planetary nebulae, the ghostly remains of dying stars. Because of its shape, this one is well named the Stingray Nebula.

ABOVE JET PROPELLED
In the giant elliptical galaxy M87 at the heart of the huge Virgo cluster of galaxies, the Hubble Telescope has spied a jet of radiation traveling at 99 percent of the speed of light. The jet emanates from a supermassive black hole in the central core of this extraordinarily active galaxy.

BELOW STARBURST SPECTACULAR
The Hubble Telescope can peer right into the heart of distant galaxies and record what is happening there. It reveals in this galaxy (NGC 4314) a bright ring, which shows that an unusual outburst of star formation is taking place close to the galaxy's nucleus (center).

How Hubble Works

With a length of 43 feet (13 m) and diameter of 14 feet (4.3 m), the Hubble Space Telescope (HST) is a big satellite. It weighs over 12 tons (11 tonnes). It was designed with grapple fixtures and hand-holds so that it could be serviced by astronauts in orbit. Electricity to power both its instruments and radio is provided by twin panels of solar cells.

The HST comprises a reflecting telescope and instruments to detect and study the light it captures. The primary light-gathering mirror is 95 inches (2.4 m) across.

Focused light from the telescope is fed to cameras and other science instruments located behind the primary mirror. The original instruments have all been replaced during a number of servicing missions. The most widely used of them was the Wide Field/Planetary Camera. Later instruments included the Advanced Camera for Surveys and NICMOS, a camera that "sees" in the infrared.

With further servicing missions, the HST could continue operating to about 2010. But without them, it could fail years earlier.

Plans are already well advanced for its replacement, known as the James Webb Space Telescope, which could become operational as early as 2011. With a primary mirror of about 20 feet (6 m), it will operate mainly in the infrared. And it will be placed in a solar orbit about 930,000 miles (1,500,000 km) away from Earth.

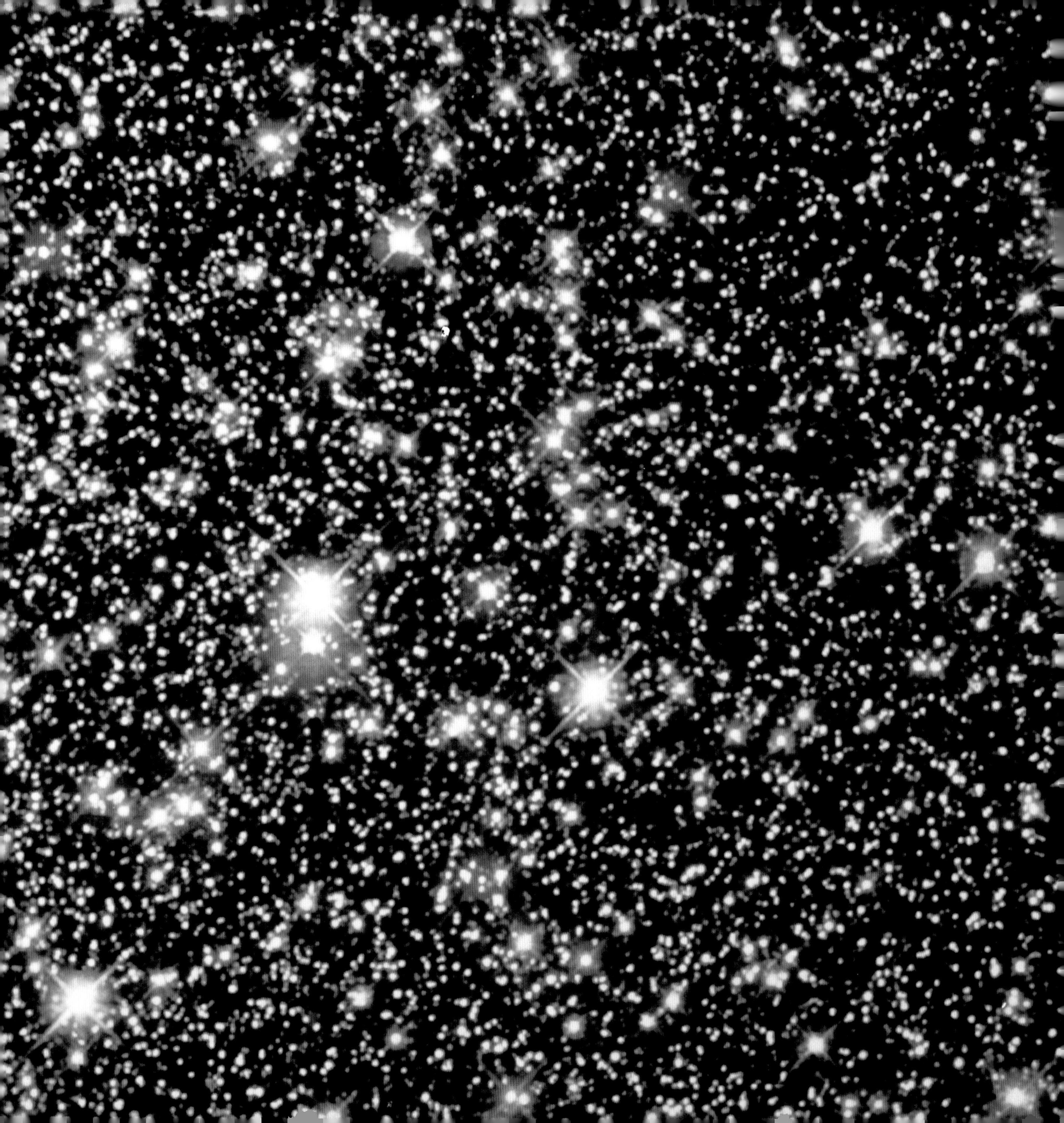

2 Looking at stars

When we look up at the night sky, we see stars by the thousands—pinpricks of light in the inky blackness of space. Millions more stars swim into view in our telescopes. And hundreds of billions of stars congregate in the great star islands in space we call the galaxies.

Stars dominate the Universe. But what exactly are they like? They are great globes of searing hot gas that generate energy in their interior and radiate it away into space. If we could travel through space to view the stars close up, we would find that they are like the great globe of searing hot gas that lies on our celestial doorstep—the Sun. The stars, then, are distant suns; or rather we should say that the Sun is a nearby star.

As stars go, the Sun is a very average kind of star, not particularly big nor particularly bright. There are billions of stars like it in the heavens. But there are also stars that are very different. There are supergiant stars of enormous dimensions, hundreds of times wider than our Sun; and there are also tiny neutron stars no bigger than a city. There are stars that blaze with the power of tens of thousands of Suns, and those that fluctuate wildly in brightness.

In recent years, thanks largely to the superlative images being sent back by new-generation telescopes on the ground and in space, astronomers have worked out exactly how stars live and die, and how they behave *en masse* in the galaxies.

LEFT CLOUD OF STARS
Flashing all colors of the rainbow, stars mass in their millions in the Milky Way in Sagittarius. This dense starscape, known as the Sagittarius star cloud, lies in the direction of the center of our Galaxy.

Distant Suns

No matter how powerful a telescope you look through, you can never see the stars other than as tiny pinpoints of light. That is because they are so far away. The stars lie at such huge distances that the ordinary units we use to measure distances on Earth, such as the mile and the kilometer, are woefully inadequate.

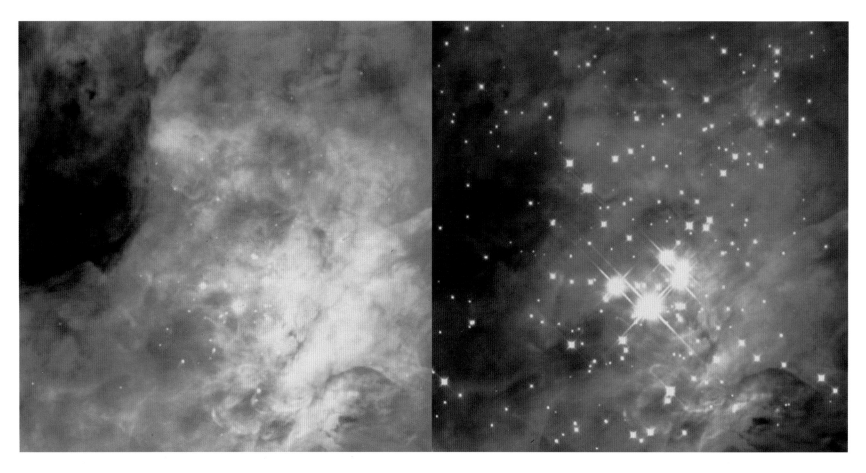

ABOVE ALL LIT UP
Young, hot stars in the Orion Nebula give off intense radiation, which makes the nebula glow. They lie at a distance of approximately 1,500 light-years. The infrared image of the same area (right) also shows a number of feebly glowing brown dwarfs—objects that never quite managed to become fully-fledged stars.

LIGHT-YEARS AWAY

To put matters into perspective, even the nearest stars lie more than 25 million million miles (40 million million km) away, a distance that is almost impossible for the human mind to grasp.

So astronomers look at things in a different way. They know that light from these nearby stars takes just over four years to reach us, so they say that the stars lie just over 4 light-years away. They are using the light-year—the distance light travels in a year (5.9 million million miles, 9.5 million million km)—as a unit of distance measurement.

In this book we express stellar distances in terms of light-years. But professional astronomers use a unit called the parsec, which is equal to about 3.3 light-years.

MAGNITUDES OF BRIGHTNESS

As a quick glance at the night sky shows, stars vary widely in brightness. Stars like Sirius, Canopus, and Rigel shine like celestial beacons, while others are scarcely visible to the naked eye.

In about 150 B.C., the Greek astronomer Hipparchus devised a scale of "magnitudes" for grading the brightness of stars in the night sky. He allotted a magnitude of 1 to the brightest stars, a magnitude of 6 to those just visible, and magnitudes 2 through 5 for stars with brightness in between.

Astronomers today still use a similar system, but have refined and extended it. On this scale, a 1st-magnitude star is exactly 100 times brighter than a 6th-magnitude star. Because brightness can now be measured accurately with instruments, astronomers express magnitudes to one or two decimal places, such as 1.25 for Deneb.

To express the brightness for stars too faint for the eye to see, the magnitude scale is extended beyond 6. Proxima Centauri, the closest star to us except for the Sun, has a magnitude of 11.1. Conversely, to express the brightness of very bright stars, the scale is extended backwards to negative values. Sirius, the brightest star in the sky, has a magnitude of -1.45.

Apparent and Absolute

The brightness of a star as we see it from Earth is only an apparent brightness. It takes no account of how far away the star is: a nearby truly dim star might appear brighter than a more distant truly bright star.

To compare true brightness, we would need to view the stars from the same distance. And this is the basis of the scale of true, or absolute, magnitude astronomers use. They define the absolute magnitude of a star as the magnitude they would observe if the star were at a distance of 10 parsecs (33 light-years). On this absolute scale, Sirius now rates only 1.4, while Rigel blazes at -7.

Variable Stars

Most of the stars we see in the sky shine steadily. (They appear to twinkle, but this is because of swirling air currents in the atmosphere.) However, some stars, named variable stars, change noticeably in brightness over a short or a long period.

Some stars vary in brightness because they are binary, or two-star systems. We refer to them as eclipsing binaries (see page 34). Other stars, nearing the end of their life, vary in brightness as they pulsate. They include the Cepheids and long-period variables like Mira. Other stars, called cataclysmic variables, change dramatically in brightness when they explode. They include novae and supernovae (see page 33).

Nearest stars	Distance (Light years)	Proper motion ("/yr)	App. magnitude	Absolute magnitude	Spectral type
Proxima Centauri	4.3	3.9	11.1	15.5	M5
α Centauri A	4.3	3.7	-0.3	4.4	G2
α Centauri B	4.3	3.7	1.3	5.7	K5
Barnard's Star	5.9	10.3	9.5	13.3	M5
Wlolf 359	7.6	4.7	13.5	16.7	M8
Lalande 21185	8.1	4.8	7.5	10.5	M2
Sirius A	8.8	1.3	-1.5	1.4	A1
Sirius B	8.8	1.3	8.7	11.6	A
Luyten 726-8A	8.9	3.4	12.5	15.3	M5
UV Ceti	8.9	3.4	13.0	15.8	M6
Ross 154	9.4	0.7	10.6	13.3	M4
Ross 248	10.3	1.6	12.3	14.8	M6
ε Eridani	10.8	1.0	3.7	6.1	K2
Luyten 789-6	10.8	3.3	12.2	14.6	M7
Ross 128	10.8	1.4	11.1	13.5	M5

Brightest stars	Designation	App. magnitude	Absolute magnitude	Spectral type	Distance (Light years)
Sirius	α Canis Majoris	-1.5	1.4	A1	8.8
Canopus	α Carinae	-0.7	1.4	F0	196.0
Rigil Kent	α Centauri	-0.3	-4.7	G2	4.3
Arcturus	α Boötis	-0.1	4.4	K1	37.0
Vega	α Lyrae	0.0	-0.2	A0	26.0
Capella	α Aurigae	0.1	0.5	G8	46.0
Rigel	β Orionis	0.1	-0.6	B8	815.0
Procyon	α Canis Minoris	0.4	-7.0	F5	11.0
Achernar	α Eridani	0.5	2.7	B5	127.0
Hadar	β Centauri	0.6	-2.2	B1	391.0
Altair	α Aquilae	0.8	-5.0	A7	16.0
Betelgeux	α Orionis	0.8	2.3	M2	652.0
Aldebaran	α Tauri	0.8	-6.0	K5	68.0
Acrux	α Crucis	0.9	-0.7	B1	261
Spica	α Virginis	1.0	-3.5	B1	261

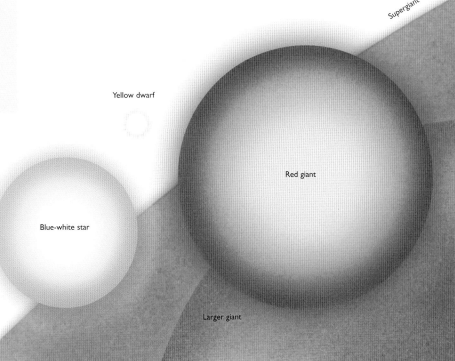

RIGHT **Dwarfs and Giants**
The difference in size among the stars is enormous. With a diameter of less than 1 million miles (1.6 million km), the Sun is regarded as a dwarf. Red giants are tens of times larger, and supergiants much larger still. Some supergiants have a diameter of hundreds of millions of miles.

The Message in Starlight

Although the stars lie very far away, we know a great deal about them. We can often tell how hot they are, what they are made of, how fast they are traveling, and so on. All this information—and more—can be extracted from the feeble rays of light that reach us through space. We do this by passing the light through a spectroscope, the most important astronomical instrument after the telescope.

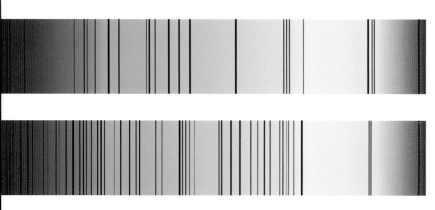

ABOVE **STELLAR SPECTRA**
The spectra of two different kinds of stars. The top spectrum is typical of white stars like Deneb, with a surface temperature of about 20,000°F (11,000°C). The bottom spectrum is typical of cool, red stars like Betelgeuse, with a temperature of only about 5,000°F (3,000°C).

SPLITTING UP LIGHT

The ordinary white light we are used to, from the Sun and the stars, is not really white at all. It is a mixture, literally, of all the colors of the rainbow. White sunlight is split into a rainbow of colors when it passes through raindrops in the atmosphere. These colors represent the way we see the various wavelengths of which light is made up.

We refer to the spread of colors (or wavelengths) in the rainbow as a spectrum (spectra in the plural). The colors go from violet (shortest wavelength), through indigo, blue, green, yellow, and orange, to red (longest wavelength).

In astronomy, white starlight is split into a spectrum when it is passed through a spectroscope (or spectrograph). Inspected closely, the stellar spectrum is crossed at intervals by a series of dark lines. And it is from the position and nature of these lines in the spectrum that astronomers can glean so much information.

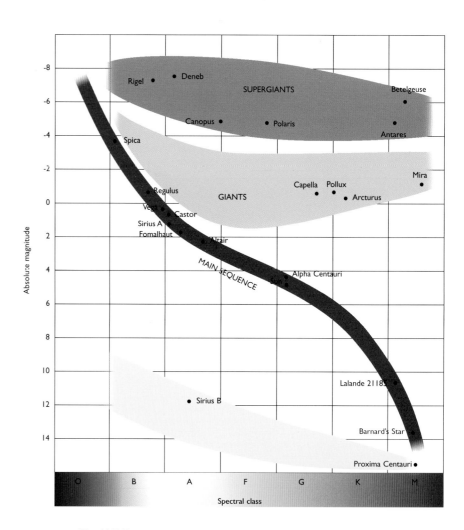

ABOVE **THE H-R DIAGRAM**
Some of the best-known stars are plotted on this version of the H-R Diagram. Most stars, including the Sun, lie on the diagonal band we call the Main Sequence. Stars remain on the Main Sequence for the majority of their lives.

THE MESSAGE IN STARLIGHT

Many different types of spectra are found among stars. For example, some are brightest in blue, some in yellow, others in red. The color intensity is a measure of their surface temperature. Spectra also differ in the number and arrangement of the dark lines. From the position of the lines, for example, we can tell the star's composition.

Astronomers find that the spectrum provides a useful way of classifying a star, because stars of a similar kind have a similar type of spectrum. They recognize ten main types of spectra, or spectral classes, designated O, B, A, F, G, K, M, R, N, and S, in order of increasing surface temperature, from about 70,000° to below 5,000°F (40,000°–3,000°C).

(This sequence of spectral classes can be remembered by the never-to-be-forgotten mnemonic "Oh, Be A Fine Guy, Kiss Me Right Now, Sweetie!")

THE H-R DIAGRAM

As well as spectral class, another fundamental characteristic of a star is its true, or absolute, magnitude (see page 25). When the two are plotted against one another on a graph, the stars fall into a number of distinct groups.

The resulting plot is called the Hertzsprung-Russell Diagram, for the astronomers who first devised it—Ejnar Hertzsprung in Denmark and Henry Norris Russell in the United States. A version of the H-R Diagram appears on the left.

RIGHT HOT STUFF
With a surface temperature of some 90,000°F (50,000°C), this Wolf-Rayet star is ten times hotter than the Sun. It is ejecting vast amounts of gas and dust into space at speeds of more than 100,000 mph (160,000 km/h). Wolf-Rayet stars are very hot and brilliant, but very short-lived.

Clouds among the Stars

We tend to think that the space between the stars is empty. But it isn't. Scattered here and there in interstellar space are traces of matter in the form of gas molecules and particles of dust. In general, this matter is very thinly distributed. But in places it clumps together into denser clouds, which we can see or detect.

LEFT ORION NEBULA
The Great Nebula in Orion, also known as M42, as viewed by the Hubble Space Telescope. The telescope has shown extraordinary detail in this nearby nebula, finding hundreds of proplyds, or solar systems in the making.

NEBULAE

Astronomers call the clouds of matter between the stars nebulae, from the Latin word for clouds. But interstellar clouds are nowhere near as dense as the clouds in our atmosphere. In fact they are a million million million times less dense even than air!

Nevertheless, nebulae do make their presence known, and we can see many of them through binoculars and telescopes because they give off light. We can even see a few with the naked eye. These visible, or bright, nebulae may be lit up in one of two ways: reflection nebulae shine because they reflect the light from nearby stars; emission nebulae shine because their gas molecules themselves emit light. Hot stars within the nebulae give off very powerful radiation that excites, or gives extra energy to, the gas molecules. The molecules then get rid of this extra energy by emitting radiation themselves.

REFLECTION AND EMISSION

The best-known bright nebula is the one we can see with the naked eye in Orion. It appears as a bright misty patch beneath the three stars that form Orion's Belt. It is a typical example of an emission nebula, being powered by the radiation of embedded hot stars.

Other emission nebulae are equally spectacular, among them the Lagoon and Trifid Nebulae in Sagittarius, the Eagle Nebula in Serpens, the Rosette Nebula in Monoceros, and the North America Nebula in Cygnus. This last nebula (see right) is a typical emission nebula, characteristically red in color. This is because red is the color of the radiation that hydrogen atoms give out, and hydrogen is the main gas found in nebulae.

Many other chemical substances are found in nebulae too—elements such as oxygen and nitrogen, together with many chemical compounds. These compounds include carbon oxides, ammonia, water, alcohols, and even simple amino acids. Amino acids are organic substances that are the building blocks of life as we know it. This suggests that life may be relatively common in the Universe.

In the Dark

There are many clouds of gas and dust in the heavens that are not lit up by nearby or embedded stars. But we can still sometimes detect them when they blot out the light from more distant stars or nebulae. To our eyes, they appear black, and are termed dark nebulae.

We can see two large dark nebulae with the naked eye. One is called the Cygnus Rift. It is found in the constellation Cygnus and looks like a hole in the dense starscape of the Milky Way. The other is found in Crux, the Southern Cross. It is called the Coal Sack. But the best-known dark nebula, visible only via telescopes, is the aptly named Horsehead (right). It really does look like a horse's head, with flowing mane.

RIGHT Horsehead Nebula
One of the most distinctive shapes in astronomy, a horse's head, which gives this dark nebula in Orion its name. The impenetrable dark matter of the nebula, also known as Barnard 33, is silhouetted against the bright nebula IC 434. This nebula extends south from Zeta Orionis, in Orion's belt.

BELOW RIGHT North America
This bright nebula in Cygnus is NGC 7000. But it is also named the North America Nebula because of its uncanny resemblance to that continent. It can be spotted in binoculars close to Deneb, but a telescope is needed to define its characteristic shape.

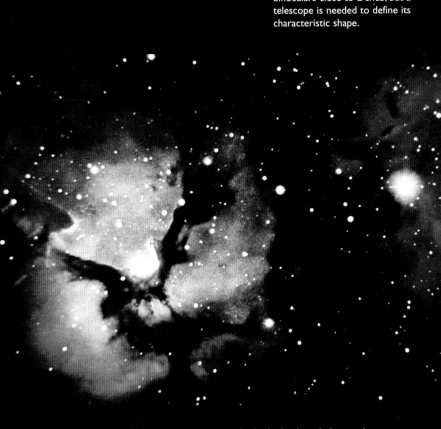

ABOVE Trifid Nebula
One of the many delightful nebulae in Sagittarius, M20. It is also named the Trifid because dark dust lanes divide the bright nebula into three parts. It is a rare combination of an emission and a reflection nebula.

Star Birth

Just like living things on Earth, stars are born, live out their lives, become old, and eventually die. But the lifespan of stars is measured in tens of millions or even tens of billions of years. Stars are born in the billowing clouds of gas and dust—nebulae—that exist in the seemingly empty space between the stars.

RIGHT **THE EAGLE'S LAIR**
The top of one of several towering pillars of gas in the Eagle Nebula in Serpens. Inside the pillars are a number of new stars, which will be revealed when the pillars are evaporated by powerful radiation from hot new stars in the background.

COLLAPSING CLOUDS

Star formation begins in the densest and coolest regions of dark nebulae in what are called giant molecular clouds.

Typically, these clouds are still and at a temperature of around -430°F (-260°C). Under these conditions, matter exists as molecules rather than as individual atoms. That is why these regions are named molecular clouds. As in other nebulae, the commonest molecule is hydrogen.

Such clouds may remain quiescent for millions of years until something disturbs them, such as the shockwave from a nearby supernova explosion. Such an event makes parts of the cloud so dense that gravity—the mutual attraction between particles—becomes the dominant force. And this causes the cloud to collapse.

CLUMPING TOGETHER

As the molecular cloud collapses, it fragments into smaller and smaller clumps, which are the starting points for individual stars to grow.

As each clump collapses under gravity, its core heats up. And as the process continues, temperatures rise to millions of degrees. The collapsing mass is now well on the way to becoming a star—we refer to it as a protostar. By now the protostar is spinning furiously, and a flattened disk of gas and dust has formed around it.

When temperatures inside the protostar reach about 18 million°F (10 million°C), the nuclei (centers) of the hydrogen atoms are forced to fuse (join) together, forming a new element, helium. In this nuclear fusion process, fantastic amounts of energy are given out, which travels to the surface and is radiated into space. A new star is born.

STEADY AS SHE GOES

Until the time nuclear reactions begin, the star has been collapsing under gravity. Now the pressure of radiation leaving the surface starts pushing against infalling matter. Eventually, radiation pressure outward just balances the fall of matter inward, and the star reaches a steady, stable state.

It will remain in this state for many millions or even billions of years, shining steadily. It becomes a main sequence star, found on the diagonal band of the H-R Diagram (see page 27).

BELOW **DARK GLOBULES**
Dark globules of gas and dust just a few light-years across are silhouetted against a bright background of radiation given off by newborn hot, massive stars. Each of these globules contain enough mass to spawn up to about ten stars.

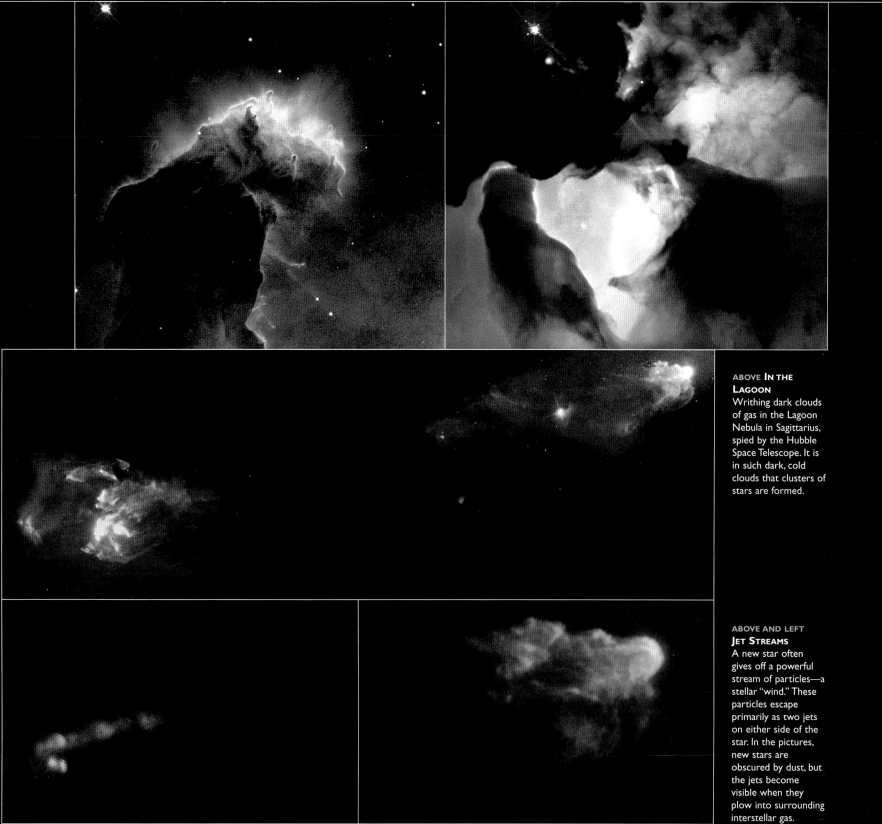

ABOVE IN THE LAGOON
Writhing dark clouds of gas in the Lagoon Nebula in Sagittarius, spied by the Hubble Space Telescope. It is in such dark, cold clouds that clusters of stars are formed.

ABOVE AND LEFT JET STREAMS
A new star often gives off a powerful stream of particles—a stellar "wind." These particles escape primarily as two jets on either side of the star. In the pictures, new stars are obscured by dust, but the jets become visible when they plow into surrounding interstellar gas.

Star Death

The lifespan of a star depends largely on its mass. Stars about the same size as the Sun will live for many billions of years, and then die relatively quietly. But stars that are much more massive may live for only a few million years, and end their lives violently.

RIGHT **SUPER SUPERNOVA** In February 1987, a supergiant star named Sanduleak 69°202 exploded near the edge of the Tarantula Nebula in our neighboring galaxy, the Large Magellanic Cloud. The exploded star is located inside the circle in the middle of the picture which was taken by the Hubble Space Telescope in 1990.

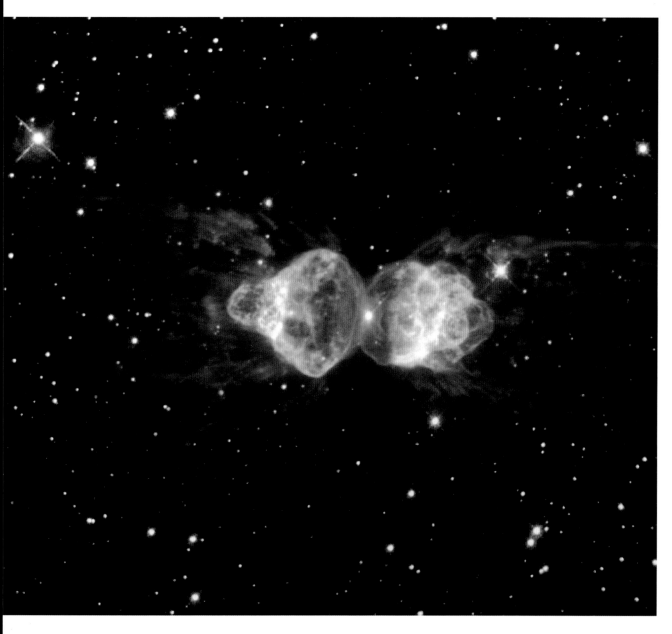

BECOMING GIANTS

A star spends most of its life shining steadily, producing energy in its core by the nuclear fusion of hydrogen into helium. But eventually its supply of hydrogen runs out, leaving just helium in the core. Now, inexorably, the star begins to die.

A relatively small star like the Sun shines steadily for around 10 billion years. Then, with the hydrogen fuel exhausted, its core collapses and heats up, and makes the outer atmosphere expand greatly. The star grows to 30 or more times its original size, becoming a red giant.

New nuclear reactions begin in the now hotter core that transforms the helium into heavier elements. This keeps the star shining for a further two billion years or so. Then, the helium runs out, and no further nuclear reactions can take place.

END GAME

The star's core shrinks even smaller and becomes superhot, with a surface temperature as high as 180,000°F (100,000°C). We now refer to it as a white dwarf. It puffs off the original star's outer layers into space as expanding shells, which glow as they push up against surrounding interstellar gas.

LEFT **THE ANT** This extraordinary object is the planetary nebula Mz3, otherwise known as the Ant because of its amazing shape. The dying white dwarf star at the center has been ejecting gas for thousands of years. Quite why the gas clouds have formed such an intricate structure is baffling. It might be something to do with the star's powerful magnetic field.

STAR DEATH

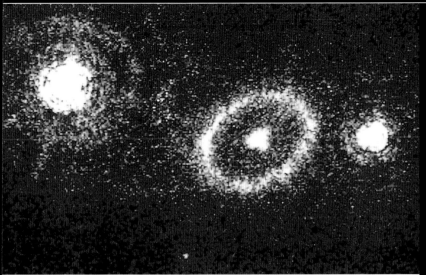

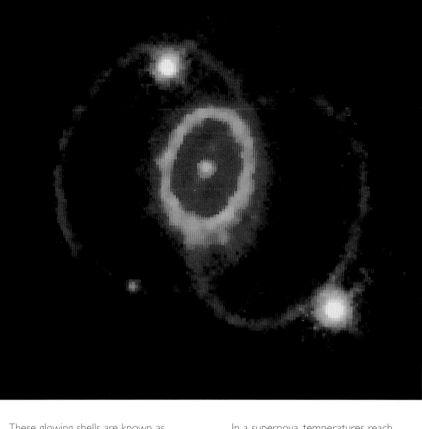

BELOW **CASSIOPEIA CALLING**
This image is a radio picture of Cassiopeia A, a powerful radio source in the constellation Cassiopeia. It is the remnants of a supernova that exploded probably in the late 1600s. Very faint visually, Cas-A is one of the brightest objects in the heavens at radio wavelengths.

RIGHT **RING A RING**
A system of multiple rings now surrounds the star that exploded as the 1987A supernova. The rings mark where shock waves and debris from the explosion plow into the surrounding gas of the interstellar medium and make it glow. This Hubble picture was taken in 1994.

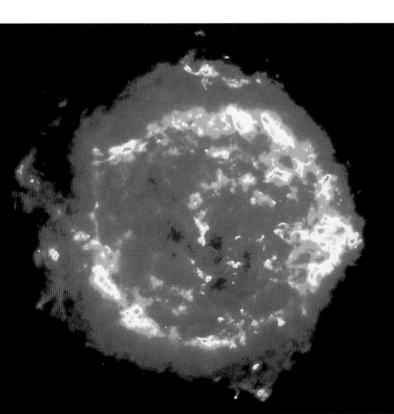

These glowing shells are known as planetary nebulae. They were so called because in telescopes they look rather like the disks of planets. Planetary nebulae have proved to be some of the most beautiful objects in the Universe when seen through the eyes of the Hubble Space Telescope. Many look like vast celestial smoke rings, like the Ring Nebula. Others have much more complex shapes, like the Ant Nebula (left), the Cat's-Eye Nebula, and the Eskimo Nebula.

SUPERGIANTS, SUPERNOVAE

A star much more massive than the Sun suffers a more spectacular fate. When it begins to die, it expands beyond the red giant stage to become a supergiant, with hundreds of times the diameter of the Sun. The supergiant is unstable, and its core soon collapses. The rest of the star collapses as well, releasing enormous gravitational energy, which blows the star to bits. We call this cataclysmic event a supernova.

In a supernova, temperatures reach hundreds of millions of degrees and pressures are colossal. Under these conditions, nuclear reactions take place that forge the host of chemical elements found in nature. These are then scattered into space and find their way into nebulae, which eventually spawn new stars.

The outcome of a supernova explosion depends on the mass of the remaining core. If it has up to about three times the mass of the Sun, it collapses to become a very dense body—a neutron star. Made up of solid neutrons, this star measures typically only about 12–20 miles (20–30 km) across.

When the core is even more massive, it collapses beyond the neutron-star stage, and is crushed virtually out of existence. All that remains is a region of space with such enormous gravity that nothing—not even light—can escape from it. It becomes that most awesome of astronomical bodies—a black hole.

Star Groups and Clusters

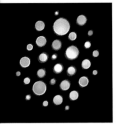

The Sun is a star that travels through space alone. But most stars travel along with one or more companions. Out of every 100 stars, about 30 lead solitary lives, about 50 are binary (two-star) systems, and the remainder are multiple stars, with three or more components (individual stars).

BELOW **THE ARCHES**
The Hubble Space Telescope has peered deep into the center of our Galaxy to spot this compact star cluster, the Arches. Only about two million years old, it contains upward of 10,000 stars, 100 times more than in the young clusters found in our Galaxy's spiral arms.

ABOVE **THE SEVEN SISTERS**
The Pleiades star cluster is also called the Seven Sisters because people with keen eyesight might be able to see its seven main stars. Brightest among them is Alcyone (in the center of the picture), which is just above the 3rd-magnitude.

DOUBLES AND BINARIES

One of the best-known patterns of stars in the Northern Hemisphere is the one we call the Big Dipper (or Plow) in the constellation Ursa Major. When you look carefully at Mizar, one of the stars in the "handle," you can see it has a close companion, named Alcor.

We refer to this pair, Mizar-Alcor, as a double star. But the two stars are not actually close together—they just happen to both lie in the same direction as we view them. Such double stars are known as optical doubles.

But other pairs of stars are really close together, bound by their mutual gravity. These are binary stars. A small telescope, for example, will reveal that Mizar is a binary, with a 4th-magnitude companion.

THE WINKING DEMON

In the constellation Perseus, the star Beta is named Algol, or the Winking Demon. It is one kind of variable star. Every 69 hours it dims for a while from about the 2nd- to the 4th-magnitude before regaining its former brilliance.

Algol behaves as it does because it is a binary star, with components orbiting in our line of sight. The light from the system dims periodically when each star temporarily passes in front of, or eclipses, the other. We know of many variable stars like this, and we call them eclipsing binaries.

STAR GROUPS AND CLUSTERS

Open Clusters

Stars also group together on a larger scale to form open clusters. The outstanding example of an open cluster is the Pleiades in Taurus. In this cluster, about 100 stars are loosely grouped together in a region about 15 light-years across. The stars are hot, blue, and comparatively young—only about 80 million years old.

Also in Taurus is another loose cluster, named the Hyades. The stars form a rough V-shape around the brilliant orange Aldebaran, although this star is not part of the cluster.

Globular Clusters

Stars also group together on an even larger scale to form great globe-shaped masses, which we refer to as globular clusters. They contain hundreds of thousands of stars, most of which are comparatively old. And whereas open clusters are to be found amongst the stars in the spiral arms of our Galaxy, globulars are found around the Galaxy's center. They follow independent orbits that often take them high above and below the general plane of the Galaxy.

Only a few of the 150 or so globular clusters we know in our Galaxy are bright enough to be seen with the naked eye. They include Omega Centauri and 47 Tucanae in the Southern Hemisphere and M13 in Hercules in the Northern Hemisphere. But many others are visible via binoculars and small telescopes. The Hubble Space Telescope has also observed many globulars in other galaxies.

ABOVE GREAT GLOBULAR The glorious globular cluster M80 in Scorpius, resolved into a colorful mosaic of thousands of individual stars by the Hubble Space Telescope. The stars are mainly old and yellowish, as to be expected, but there are also a number of young blue stars, nicknamed "blue stragglers." They might have formed when older stars collided.

The Milky Way

On clear, dark nights, you can see a hazy band of light spanning the dome of the heavens. The ancient Greeks called it the Milky Circle; we know it as the Milky Way. Binoculars and small telescopes show it to be made up of millions of stars, seemingly closely packed together.

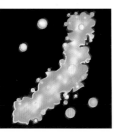

ABOVE
IN CROSS-SECTION
A map of the night sky based on data returned by the infrared astronomy satellite IRAS. It shows, in effect, a cross-section of our Galaxy. The bright band in the middle is the Milky Way. The large and small white patches underneath are our closest galactic neighbors, the Large and Small Magellanic Clouds.

THE STARRY DISK

The Milky Way passes through some of the most brilliant constellations: Cassiopeia, Cygnus, Perseus, Auriga, and Aquila in the Northern Hemisphere; and Puppis, Vela, Carina, Crux, Centaurus, Scorpius, and Sagittarius in the Southern. It is particularly bright in Scorpius and Sagittarius.

The width of the Milky Way varies greatly. In places it is only about 5 degrees across, while in others, it approaches 30 degrees. Dark rifts also appear in parts of the Milky Way—for example, in Cygnus and Aquila. They are actually bands of dust that blot out the light from stars beyond.

What exactly is this band of close-packed stars that we call the Milky Way? It represents a cross-section of the star system, or galaxy, to which all the stars we see in the night sky belong. We also call this star system the Milky Way Galaxy, or often just "the Galaxy."

SHAPING UP

The Galaxy takes the form of a disk with a bulge at the center (the nucleus). From this bulge a number of arms curve out, carrying the stars. The whole Galaxy rotates and, viewed from afar, would look rather like a flaming Catherine wheel firework.

Overall, the Galaxy has a spiral structure and is classed as a spiral galaxy (see diagram, top right). Its size is staggering—it measures 100,000 light-years across and contains as many as 200 billion stars. The Sun lies on one of the spiral arms about 30,000 light-years from the center. The center lies in the direction of Sagittarius, which is why the

ABOVE STUNNING STARSCAPE
A view of the star-studded Milky Way in the far southern constellation Norma. The dark regions show where dust is obscuring the light from stars beyond. The picture, taken in March 1986, shows a celestial interloper, Halley's Comet, which was best seen in the Southern Hemisphere that year.

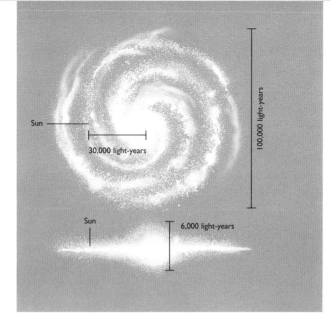

LEFT DIMENSIONS
With a diameter of around 100,000 light-years, the Milky Way Galaxy is bigger than many spirals.

Milky Way is densest and brightest there and in the adjacent constellations.

IN THE NUCLEUS

The central bulge of the Galaxy measures about 6,000 light-years across. It contains mostly old red and yellow stars. At the center of the bulge is a powerful radio source known as Sagittarius A.

Astronomers reckon that a supermassive black hole is responsible for producing this emission. Supermassive black holes, with the mass of millions of Suns, are thought to reside in the center of most galaxies.

Circling the bulge are many globular clusters. They are among the oldest objects in the Galaxy, around 10 billion years old.

IN THE SPIRAL ARMS

The spiral arms contain younger stars than the bulge, and much interstellar gas and dust, from which new stars will be born.

Investigations using radio waves to penetrate the thick dust clouds in the Milky Way have revealed the spiral structure of the Galaxy. There are two main spiral arms—Sagittarius and Perseus—and parts of others. The Sun lies on the Orion Arm, which links the two main arms.

BELOW SPIRAL STRUCTURE
Viewed from far away in space, our Galaxy would look much like this one (NGC 4414) in the constellation Coma Berenices. In the center is the nuclear bulge, containing older stars. Younger, bluer stars are found in the spiral arms that curve out from the central nucleus.

Our Galactic Neighbors

In far southern skies, you can see two misty patches, looking rather like clouds in our atmosphere. We refer to them as the Large and Small Magellanic Clouds. These are not Earthly clouds a few miles up in the sky, but separate star systems, many thousands of light-years away.

LEFT THE GREAT SPIRAL
The Great Spiral in Andromeda, otherwise known as the Andromeda Galaxy or M31. It was the first star system proved to be outside our own Galaxy. U.S. astronomer Edwin Hubble made this discovery in 1923. This picture also shows its two satellite galaxies, M32 (close in) and NGC 205.

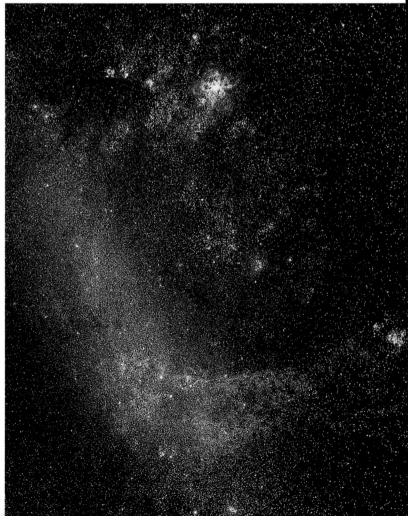

BELOW CLOSE CLOUD
The Large Magellanic Cloud (LMC) in Dorado is one of the great delights of far southern skies. Irregular in shape, it is only about one-third of the size of our own Galaxy. Near the top of the picture, the bright cloud is the Tarantula Nebula, a huge star-forming region.

MAGELLAN'S CLOUDS

The Magellanic Clouds are named for the Portuguese navigator Ferdinand Magellan, who was one of the first Europeans to sail the southern oceans, in the early 1500s. No doubt he would have used the Clouds to help him navigate.

The Large Magellanic Cloud (LMC) is located in the constellation Dorado. Some 30,000 light-years across, it lies about 160,000 light-years away. It contains much the same mix of stars, clusters, nebulae, and so on, as our own Galaxy, but has no particular structure. It is classed as an irregular galaxy.

The Small Magellanic Cloud (SMC), in Tucana, is also irregular. It is only about two-thirds the size of the LMC and lies some 30,000 light-years farther away.

The LMC and the SMC are the nearest galaxies we can see in the sky. But there is a closer one, obscured by dust in the Milky Way. Named the Sagittarius dwarf galaxy, it was only discovered late in the twentieth century. It lies only 78,000 light-years away.

The Magellanic Clouds and the Sagittarius dwarf galaxy are not only close neighbors of the Milky Way, but they circle around it, bound by gravity. They are known as satellite galaxies.

OUR GALACTIC NEIGHBORS

The Andromeda Galaxy

There is only one other galaxy that you can see from Earth, in the constellation Andromeda. It is the Andromeda Galaxy, M31. It is visible to the naked eye as a faint misty patch quite close to the star Nu (ν) Andromedae.

Like the Magellanic Clouds, the Andromeda Galaxy is one of our galactic neighbors. But it lies much farther away than they do, at a distance of about 2.5 million light-years. It is the farthest object we can see in the heavens with the naked eye.

ABOVE THE TARANTULA
The Tarantula Nebula pictured in close up by the Hubble Space telescope. It is one of the biggest and brightest of all nebulae. Around 80 light-years across, it is 50 times bigger than the Orion Nebula in our Galaxy.

One reason why we can see it at such a distance is that it is so big. It is about half as big again as our own Galaxy and contains as many as 400 billion stars. Like our Galaxy, it is a spiral. Also like our Galaxy, it has satellite galaxies, M32 and NGC 205, which are shown in the picture (opposite right).

The Local Group

Our Galaxy, Andromeda, and the Magellanic Clouds are all bound together loosely by gravity. They form part of a so-called Local Group of galaxies that travel through space together.

In all, there are about 30 galaxies in the Local Group. There are just three spirals—Andromeda, our own Galaxy, and the smaller Triangulum Galaxy, M33. All the other galaxies are much smaller, classed as dwarf irregular or dwarf elliptical galaxies according to their shape.

Galaxies Galore

In their telescopes, astronomers can see galaxies scattered throughout space in every direction. Most are either pinwheel-like spirals, similar to our own Galaxy, or ellipticals, round, or oval in shape. The others, irregulars, lack any definite structure.

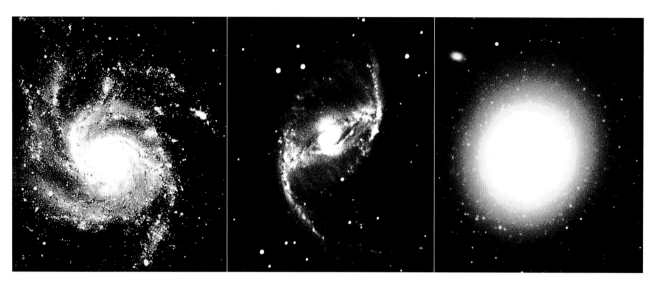

FAR LEFT SPIRAL
Galaxy M101 (NGC 5457) in Ursa Major is a classic spiral galaxy of the Sc type, with wide-open and well-defined arms. One of the biggest spirals known, it is twice as wide as our own Galaxy.

CENTER BARRED SPIRAL
Galaxy NGC 1530 is a typical barred-spiral galaxy, with a pronounced "bar" of stars running through the center of the nuclear bulge. It is found in the faint constellation Camelopardalis.

LEFT ELLIPTICAL
Galaxy M87 (NGC 4486) is an elliptical galaxy of the E1 type, being almost spherical. Found at the heart of the Virgo Cluster of galaxies, it is an active galaxy that is a powerful radio source.

Hubble's Galaxies

U.S. astronomer Edwin Hubble introduced the system we use for classifying the galaxies early in the twentieth century. He was the great pioneer investigator of other galaxies—which were once thought to be nebulae in our own Galaxy.

In 1919 he began studying these "nebulae" with the newly built 100-inch (2.5-m) Hooker Telescope at Mount Wilson Observatory. In 1923, he discovered that the Great Spiral "nebula" in Andromeda lay far beyond the confines of our own Galaxy and was a separate star system. He found that other "spiral nebulae" were similarly remote star systems in their own right. He named them extragalactic nebulae, but we refer to them as galaxies.

Hubble later discovered that all the galaxies, except nearby ones, were rushing headlong through space away from one another. And the farther they were away, the faster they were traveling. This suggested that the Universe was expanding.

It was thus entirely appropriate that the Hubble Space Telescope—designed to "open up a new window on the Universe"—should be named after him. He would have reveled in the images the telescope has been sending back.

Classifying the Galaxies

The majority of galaxies have a definite shape—known as regular galaxies. Others have no definite shape or structure and are irregular.

We classify the regular galaxies (after Hubble) according to their shape. Spiral galaxies (S) are graded a, b, or c, according to the openness of their spiral arms. Sc galaxies have wide-open arms. Some spiral galaxies have a kind of bar through their center and are called barred-spiral galaxies (SB). They too are graded a, b, or c, according to how open their arms are.

The Andromeda Galaxy has moderately open spiral arms and is classed as an Sb. Traditionally, our own Galaxy has been regarded as an Sb too, but there is some evidence of a slight bar through the nucleus, so perhaps it should be classed as an SBb.

Elliptical galaxies (E) are graded 0-7 according to how spherical or oval they are. E0 galaxies are spherical, E7 the most flattened ovals.

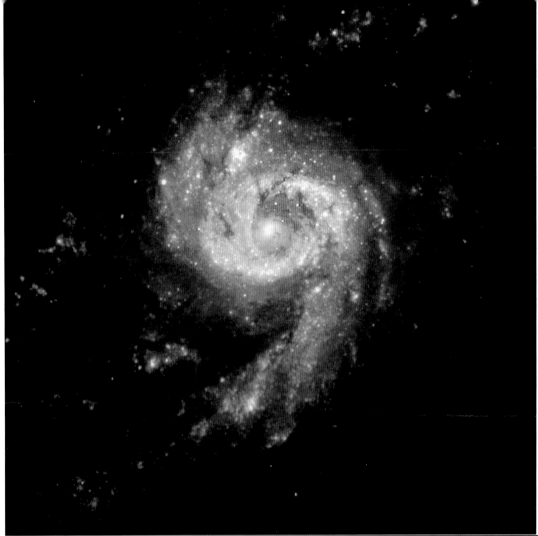

RIGHT STARBURST GALAXY
Hubble Space Telescope image of NGC 3310, which shows an unusually active star formation. It is a type known as a "starburst" galaxy, in which clusters of new stars are forming at a prodigious rate, their youth indicated by their blue color.

What Galaxies Are Like

Spiral galaxies have much the same structure as our own Galaxy—a central bulge of older stars with dusty spiral arms coming out from it carrying younger stars. Very old globular clusters circle independently around the bulge.

Elliptical galaxies lack the arms of spirals and are made up mainly of older stars. There is little free gas and dust present to trigger new star formation. Over half of all galaxies seem to be ellipticals. They differ widely in size, from dwarf ellipticals containing just a few million stars to giant ellipticals containing hundreds of billions of stars. They vary in size from a few hundred light-years across to hundreds of thousands.

RIGHT WARPED GALAXY
Edge-on view of the spiral galaxy ESO 510-G13. Dark dust lanes are visible along the plane of the disk, as is usual in spirals. Unusually, however, the disk is distorted. Astronomers reckon that this happened as a result of a collision with another galaxy long ago.

Hyperactive Galaxies

Most galaxies give off the amount of energy you would expect from a collection of hundreds of billions of stars. But some galaxies give out exceptional energy either as visible light or at invisible wavelengths, such as X-rays. Astronomers name them active galaxies.

The Tale of 3C-273

Radio astronomers often identify radio emissions in the heavens as a number in the third Cambridge catalog (3C) of radio sources. In 1963, radio astronomers were observing one named 3C-273 when it was occulted (covered) by the Moon. They identified it visually with a star and took its spectrum.

Astonishingly they found from the red shift in the spectral lines that it was more than 2 billion light-years away. 3C-273 was clearly no ordinary star. To be visible at such a distance, it had to be as bright as hundreds of galaxies put together. Astronomers named this strange new object a quasar, standing for "quasi-stellar radio source."

Active Galaxies

Since that time hundreds of other quasars have been discovered, pumping out exceptional energy not only as radio waves and visible light, but also as X-rays and infrared rays. They prove to be just one type of active galaxy. Other kinds of active galaxies include radio galaxies, Seyfert galaxies, and blazars.

Radio galaxies are notable for their exceptional emission of radio waves. Centaurus A and M87 in Virgo are two of the most powerful radio galaxies. In general, a radio galaxy emits most of its radio energy from regions, known as lobes, on either side of it. These lobes may span distances of millions of light-years.

Seyfert galaxies are active spiral galaxies noted for their exceptionally brilliant center. They are named after U.S. astronomer Carl Seyfert, who first investigated such galaxies in the early 1940s.

LEFT QUASI-STELLAR
Billions of light-years away, this quasar looks star-like in ground-based telescopes. But the sharp eyes of the Hubble Space Telescope reveal it to be the brilliant center of a spiral galaxy.

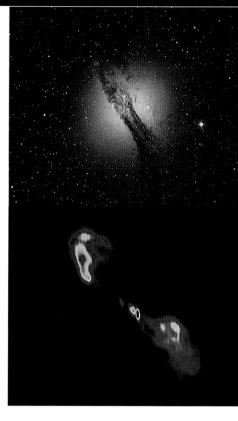

TOP VISIBLE IMAGE
The galaxy NGC 5128 in Centaurus, also called Centaurus A, is readily spotted in binoculars. Telescopes will show it is bisected by a dark band of obscuring dust.

ABOVE RADIO IMAGE
Radio telescopes reveal that Centaurus A is a radio galaxy—in fact, the third most powerful radio source in the heavens. The galaxy itself is located in the center of this image. Most radiation comes from the two "lobes" on either side.

Quasars tend to vary in brightness, sometimes on a daily basis. So do another class of active galaxy known as blazars. Blazars are named after the object BL Lacertae, which varies widely and rapidly in brightness. It was once thought to be a kind of variable star.

LEFT JET PROPELLED
The Hubble Space Telescope view of the heart of the giant elliptical galaxy M87 in Virgo. Enlarged is the central core, the location of a supermassive black hole that is responsible for the prodigious amounts of energy this active galaxy puts out, particularly as X-rays and radio waves. Also visible is a jet of electrons coming from the black hole and traveling at nearly the speed of light.

THE ENERGY MACHINE

All the active galaxies radiate their exceptional energy from their central region, over an area roughly the size of the Solar System. Only one known power source could generate the power they do—a supermassive black hole.

"Ordinary" black holes form after the biggest stars die and blast themselves apart in a supernova explosion. They have the mass of several Suns. But the supermassive black holes at the heart of active galaxies have the mass of hundreds of millions of Suns. They seem to form at the heart of galaxies by the catastrophic collapse of clouds of matter there.

Astronomers believe that the various kinds of active galaxies are really views of supermassive black holes from different angles.

RIGHT SPECTACULAR SEYFERT
A beautiful face-on view of the Seyfert galaxy NGC 7742, which has the very bright center characteristic of this type of active galaxy. The broad blue-white ring around the brilliant center is a region of intensive star birth.

Galaxies and the Universe

Galaxies are generally well scattered in space, with millions of light-years between them. But sometimes they come close enough together to interact, or even collide, with one another. Then, all kinds of celestial fireworks take place.

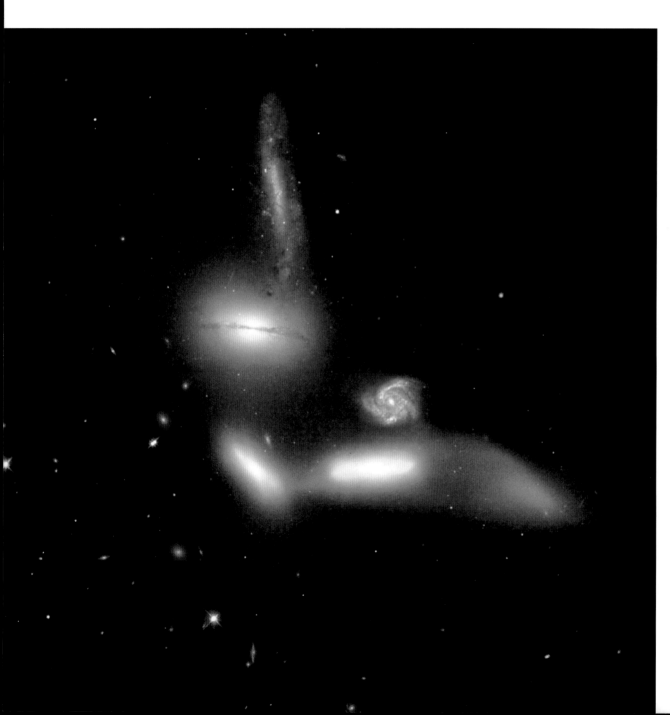

WHEN GALAXIES COLLIDE

What happens when two galaxies interact depends on what they are like to start with and how close they approach each other. What occurs is a gravitational tug of war between the two and their individual stars and clusters.

When two close spiral galaxies interact, for example, they may retain their basic structure but draw out long streamers of stars from each other. If galaxies actually collide, their structure will be distorted or destroyed. And the compression of the gas clouds in the galaxies as they slam into each other will trigger a new round of star formation, creating what are called starburst galaxies.

If one galaxy is much smaller than the other, the large one will swallow the small one with comparatively little disturbance to itself. This is happening with our own Galaxy, which is gradually absorbing the Sagittarius dwarf galaxy and is getting ready to do the same with its two satellite galaxies, the Large and Small Magellanic Clouds. In the far distant future, our own Galaxy might itself be swallowed up by the much larger Andromeda Galaxy.

LEFT GROUPING TOGETHER
U.S. astronomer Carl Seyfert was first to spot this small group of galaxies, which is now named Seyfert's Sextet. Four of the galaxies lie close together in space and interact with one another.

Clustering Together

Our Galaxy, the Andromeda Galaxy, and the Magellanic Clouds are close galactic neighbors. They form part of the 30-strong Local Group of galaxies, which spans a region of space about 5 million light-years across.

All the other galaxies also gather together in groups, or clusters. Many clusters are much bigger than the Local Group. About 50 million light-years away lies the Virgo cluster, which contains upward of 2,000 galaxies in a region of space more than 10 million light-years across. Even bigger is the Coma cluster (in Coma Berenices), containing over 3,000 galaxies in an area twice this size.

Superclusters

The many clusters of galaxies found in space form part of even bigger groupings called superclusters. Our Local Group and the Virgo cluster are part of the Virgo, or Local, supercluster. It occupies a region of space about 100 light-years across.

In a supercluster, the clusters of galaxies are linked together to form huge sheets or filaments that can stretch for hundreds of millions of light-years.

In their turn, superclusters link together throughout the depths of space to form our Universe. They tend to wrap themselves around vast empty spaces, which are known as voids. This tends to give the Universe an overall spongy texture.

BELOW CLOSE ENCOUNTER
Two spiral galaxies pass each other as they hurtle through space. The powerful gravity of the larger one (NGC 2207) is pulling out matter from the other (IC 2163), forming streamers of stars that stretch for hundreds of thousands of light-years.

ABOVE
DEEP IN SPACE
Galaxies in clusters and superclusters populate the Universe. This Hubble Space Telescope view, the result of a 120-hour exposure, shows galaxies at distances of 10 billion light-years or more. In other words, we are looking at them as they were 10 billion years ago when the Universe was probably less than 3 billion years old.

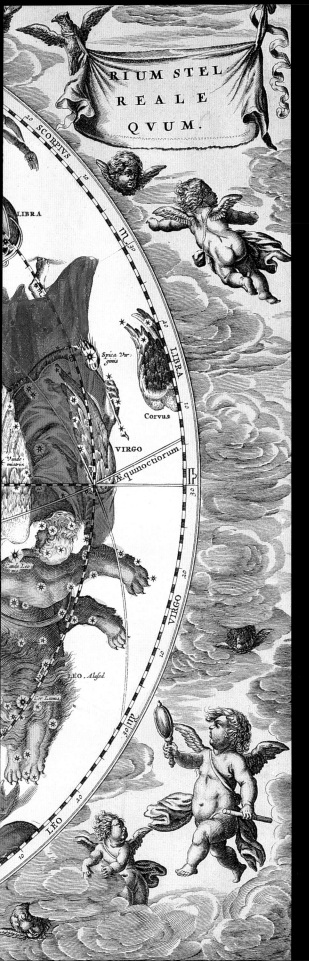

3 Patterns in the sky

Like our distant ancestors, we use the patterns made by the brightest stars in the heavens as signposts to help us find our way around the night sky. These patterns in the sky, which are known as the constellations, do not change perceptibly over the years nor indeed over the centuries.

The stargazers of the early civilizations in the Middle East, some 5,000 years ago, saw virtually the same constellation patterns we see today. The priest-astronomers of those civilizations observed the rhythms of the heavens to establish calendars that would bring a semblance of order to the lives of an expanding population in an increasingly complex society. They also looked to the heavens for signs that would indicate the will of their gods, whom they believed ruled human lives. Their "reading of the stars" led to the pseudo-science of astrology, which still has its adherents today.

Familiarizing ourselves with the constellations is the objective of this chapter. Astronomers recognize 88 constellations, many of which have fanciful names that describe the figures that the stars in that constellation represent. Because the Earth is round, observers at a certain place can see only certain constellations during the year. Other constellations remain unseen, always beneath their horizon. Since the Earth travels around the Sun once a year, constellations appear in the night sky, and disappear from it, at different times throughout the year.

LEFT NORTHERN CONSTELLATIONS
This colorful star map from the eighteenth century shows imaginative figures representing constellations visible in the Northern Hemisphere. It is bounded by the constellations of the zodiac, through which the Sun passes on its annual path around the celestial sphere.

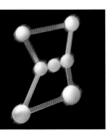

The Constellations

The patterns of bright stars in the sky—the constellations—don't appear to have changed, even after thousands of years. Their stars seem to be fixed immutably and travel as a group through space. This is, however, an illusion.

LEFT MIGHTY ORION
A splendid representation of one of the best-known constellations, Orion. The figure is depicted as a mighty hunter, with shield in his left hand and club in his right, poised ready to strike a blow. Orion was one of the great Greek heroes, son of the sea god Poseidon, who gave him the ability to walk on water.

Spaced Out

The stars in the constellations are not grouped together at all. They are quite independent of one another and are usually separated by vast distances in space. For example, of the seven brightest stars in the familiar constellation Orion, the closest (Gamma) lies 300 light-years away, while the most distant (Kappa) lies over 1,800 light-years away.

We see the stars in the constellations together in the sky simply because they happen to lie in the same direction in space as we view them from Earth.

The stars in the constellations also appear to be fixed in position, but they are not. They are all traveling independently through space at different speeds and in different directions. We don't see them move (except in a few cases) because they are so very far away. Only after many thousands of years will the star patterns we see today change noticeably.

Naming the Constellations

Altogether, astronomers recognize 88 constellations, detailed on the star maps opposite. Many of them date back at least 5,000 years and were recorded by the ancient stargazers of the Middle East, in Babylon and Egypt.

The names of many of the constellations were given by the ancient Greeks, although we now use the Latin forms of the Greek names. We also sometimes refer to the constellations by their translated names. For example, the constellation Ursa Major translates into English as the Great Bear, Leo as the Lion, and Cygnus as the Swan.

The Greeks named the constellations after gods, heroes, creatures, and objects that featured in their mythology. They matched the star patterns to mythological figures they thought the patterns resembled. But a great deal of imagination is often required to fit the one with the other.

Only in a few cases do constellations look like the figures they are meant to represent. Scorpius is an example: little imagination is required to see in the star pattern a scorpion with its deadly curved tail, poised ready to strike.

Modern Constellations

The Greeks recognized some 48 constellations, which were listed by the last great ancient Greek astronomer, Ptolemy, around A.D. 150. The other 40 constellations were added much later, notably by the German astronomers Johann Bayer (in 1603) and Johann Hevelius (in 1690), and the French astronomer Nicolas Lacaille (in 1752). Bayer also introduced the system of using letters of the Greek alphabet to identify the stars in a constellation (see page 67).

THE CONSTELLATIONS

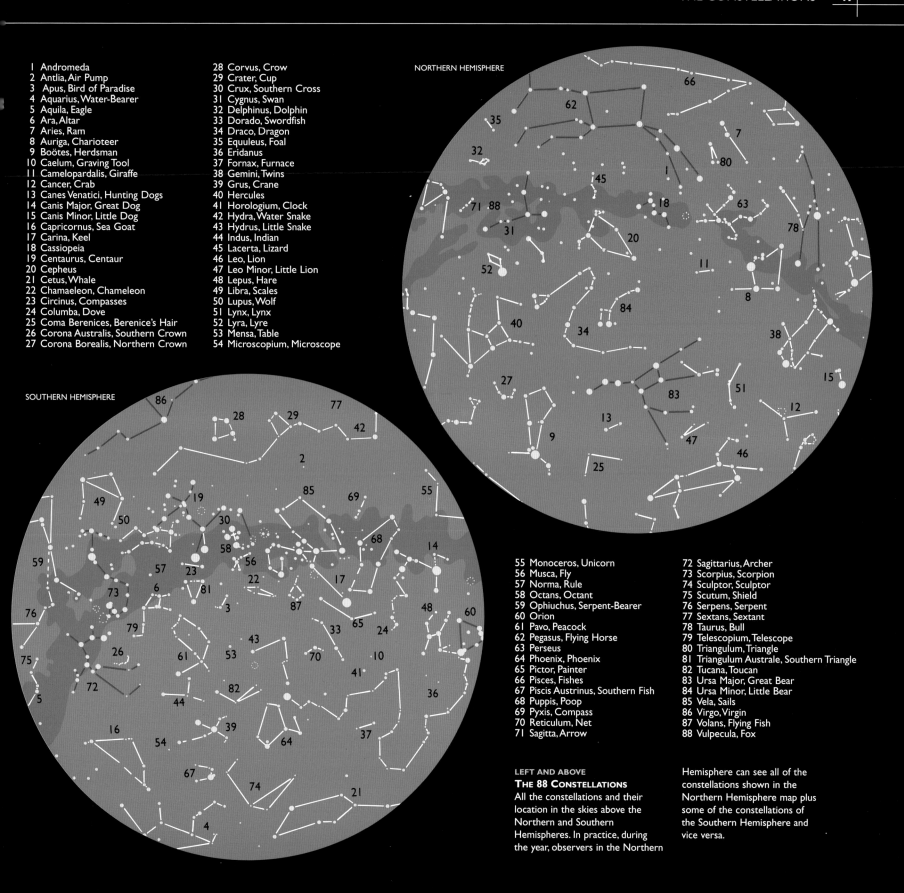

1. Andromeda
2. Antlia, Air Pump
3. Apus, Bird of Paradise
4. Aquarius, Water-Bearer
5. Aquila, Eagle
6. Ara, Altar
7. Aries, Ram
8. Auriga, Charioteer
9. Boötes, Herdsman
10. Caelum, Graving Tool
11. Camelopardalis, Giraffe
12. Cancer, Crab
13. Canes Venatici, Hunting Dogs
14. Canis Major, Great Dog
15. Canis Minor, Little Dog
16. Capricornus, Sea Goat
17. Carina, Keel
18. Cassiopeia
19. Centaurus, Centaur
20. Cepheus
21. Cetus, Whale
22. Chamaeleon, Chameleon
23. Circinus, Compasses
24. Columba, Dove
25. Coma Berenices, Berenice's Hair
26. Corona Australis, Southern Crown
27. Corona Borealis, Northern Crown
28. Corvus, Crow
29. Crater, Cup
30. Crux, Southern Cross
31. Cygnus, Swan
32. Delphinus, Dolphin
33. Dorado, Swordfish
34. Draco, Dragon
35. Equuleus, Foal
36. Eridanus
37. Fornax, Furnace
38. Gemini, Twins
39. Grus, Crane
40. Hercules
41. Horologium, Clock
42. Hydra, Water Snake
43. Hydrus, Little Snake
44. Indus, Indian
45. Lacerta, Lizard
46. Leo, Lion
47. Leo Minor, Little Lion
48. Lepus, Hare
49. Libra, Scales
50. Lupus, Wolf
51. Lynx, Lynx
52. Lyra, Lyre
53. Mensa, Table
54. Microscopium, Microscope
55. Monoceros, Unicorn
56. Musca, Fly
57. Norma, Rule
58. Octans, Octant
59. Ophiuchus, Serpent-Bearer
60. Orion
61. Pavo, Peacock
62. Pegasus, Flying Horse
63. Perseus
64. Phoenix, Phoenix
65. Pictor, Painter
66. Pisces, Fishes
67. Piscis Austrinus, Southern Fish
68. Puppis, Poop
69. Pyxis, Compass
70. Reticulum, Net
71. Sagitta, Arrow
72. Sagittarius, Archer
73. Scorpius, Scorpion
74. Sculptor, Sculptor
75. Scutum, Shield
76. Serpens, Serpent
77. Sextans, Sextant
78. Taurus, Bull
79. Telescopium, Telescope
80. Triangulum, Triangle
81. Triangulum Australe, Southern Triangle
82. Tucana, Toucan
83. Ursa Major, Great Bear
84. Ursa Minor, Little Bear
85. Vela, Sails
86. Virgo, Virgin
87. Volans, Flying Fish
88. Vulpecula, Fox

LEFT AND ABOVE
THE 88 CONSTELLATIONS
All the constellations and their location in the skies above the Northern and Southern Hemispheres. In practice, during the year, observers in the Northern Hemisphere can see all of the constellations shown in the Northern Hemisphere map plus some of the constellations of the Southern Hemisphere and vice versa.

Signposts to the Stars

In a way, every constellation helps act as a signpost to point an observer in the right direction to other stars and constellations. But some constellations are particularly valuable. They include Ursa Major in the far Northern Hemisphere, Orion on the celestial equator, and Centaurus in the far Southern Hemisphere.

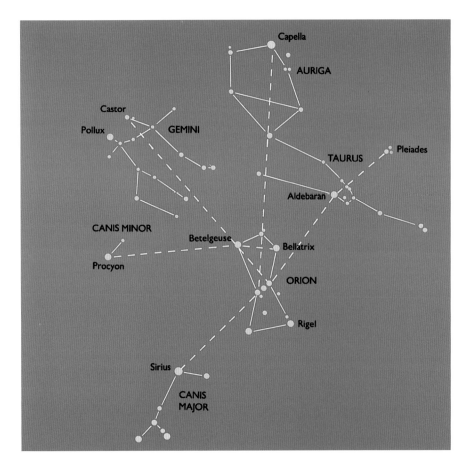

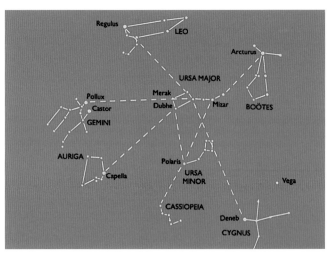

The four stars in the "cup" part of the Big Dipper act in pairs to point to other stars. A line through the best-known pair, Merak and Dubhe, points to the Pole Star, Polaris, and they are called the Pointers. Navigators have used this alignment for centuries to find their way across the ocean.

Other alignments direct you to Deneb in Cygnus, Capella in Auriga, Pollux and its twin Castor in Gemini, and Regulus in Leo. Following the curve of the dipper handle brings you first to Arcturus in Boötes and then to Spica in Virgo.

ABOVE ORION
Orion is one of the best signposts to other stars and constellations. Because it spans the celestial equator, it is equally useful for observers in both the Northern and Southern Hemispheres.

NORTHERN POINTERS

In the northern heavens, the best signpost by far is the Big Dipper (also known as the Plow), which is the most prominent part of Ursa Major. Although it boasts no 1st-magnitude stars, the Big Dipper is unmistakable. And it is circumpolar—always visible—from much of the Northern Hemisphere.

MAGNIFICENT ORION

Orion is invaluable as a signpost to observers in both the Northern and Southern Hemispheres. Its shape is distinctive and its stars are bright.

ABOVE URSA MAJOR
The Big Dipper, the prominent star pattern in Ursa Major, is an excellent signpost in the Northern Hemisphere. It can be used to locate some of the brightest stars, such as Deneb, Capella, and Arcturus.

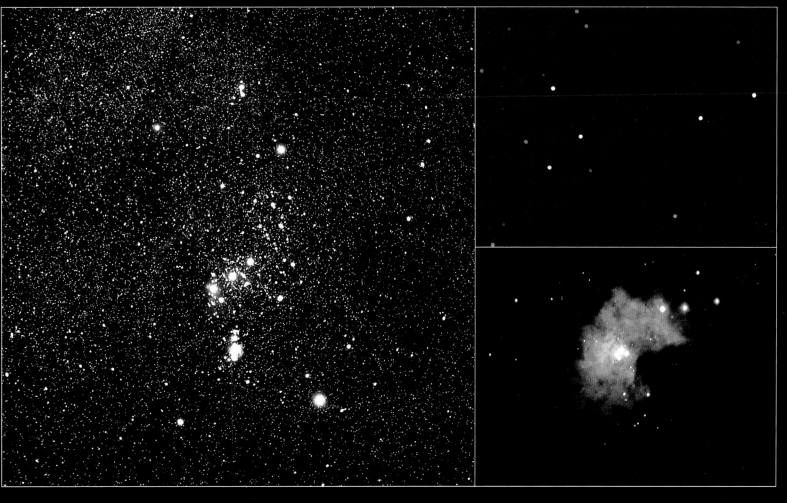

ABOVE ORION'S STARS
The distinctive pattern of bright stars that makes up the constellation of Orion. Top left is the somewhat variable 1st-magnitude star Betelgeuse, which is outshone by the star Rigel, bottom right. The bright patch beneath the three stars in Orion's Belt is the Orion Nebula.

The center of the pattern, which form Orion's Belt, are particularly useful. A line through them running south locates the brightest star in the sky, Sirius in Canis Major, also called the Dog Star.

Taking a line through them north brings you first to Aldebaran in Taurus, marking the red eye of the bull, and then to the Pleiades, or Seven Sisters, the finest open star cluster in the heavens.

Capella, Castor and Pollux, and Procyon in Canis Minor can also be found by means of other alignments in the constellation.

SOUTHERN POINTERS

There is also a notable pair of pointers in far southern skies—the 1st-magnitude Alpha and Beta Centauri in Centaurus. This pair guide the eye to the most famous of southern constellations, Crux, the Southern Cross. Without them, the eye might be tempted to stray to the False Cross nearby, formed by one pair of stars from Vela and another from Carina.

Unlike the Northern Hemisphere, the Southern Hemisphere has no convenient pole star to mark the position of the celestial South Pole, nor even pointers to it. But the long axis of the Southern Cross points nearly in the right direction.

TOP LOCATING THE CROSS
The twin bright stars Alpha and Beta Centauri (right) act as invaluable pointers to the far southern constellation, Crux, the Southern Cross.

ABOVE THE TRAPEZIUM
One of the many distinctive features of the signpost constellation Orion is the Great Nebula M42. Embedded within it is a multiple star, known as the Trapezium after the shape made by the star's four components.

The Celestial Sphere

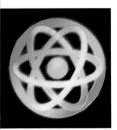

Viewed from Earth, the star-filled heavens seem to be like a great dark dome above our heads. The stars appear to be stuck on the inside of it. Ancient astronomers named this great Earth-enveloping sphere the celestial sphere.

We know today that there is no such thing as a celestial sphere: the stars all lie at different distances away from Earth, and dark space goes on, seemingly for ever.

Nevertheless, the idea of a celestial sphere is still useful in observational astronomy: in particular, we can use the geometry of a sphere to pinpoint the positions of the stars in the sky.

Features of the Sphere

Here are some significant features of the celestial sphere that astronomers frequently refer to—

The *north and south celestial poles* are points on the celestial sphere that lie directly above Earth's North and South geographical poles.

The *celestial equator* is a circle around the celestial sphere halfway between the north and south celestial poles. It is a projection of Earth's Equator onto the sphere.

An observer's view of the celestial sphere is bounded all around by the *horizon*, which is where the horizontal plane through the observer's position meets the sphere.

The point on the celestial sphere directly above the observer's head is the *zenith*. The equivalent point on the sphere directly beneath the observer is the *nadir*.

The great circle passing through the north celestial pole, the zenith, and the south celestial pole is the *meridian*. Stars *culminate*, or reach, their highest altitude in the sky on the meridian.

The *ecliptic* is the apparent path of the Sun around the celestial sphere during the year. The Sun crosses the celestial equator on about March 21 every year, moving north; and on about September 23, moving south. On these dates, day and night are exactly 12 hours long all over the world. That is why they are called the equinoxes—"equal nights."

Celestial Time

The spinning of Earth on its axis provides the basis for measuring time. Our day of 24 hours is the time it takes Earth to spin around once and return to the same point in the sky relative to the Sun. We refer to this as our ordinary time, "solar time."

We might expect the celestial sphere to rotate around us once every 24 hours, but it doesn't. If you note the time when a certain

RIGHT **HEAVENLY DOME**
The ancient concept of the celestial sphere. The stars are stuck on the inside of a huge sphere in fixed positions. Earth is stationary at the center of the sphere that rotates from east to west.

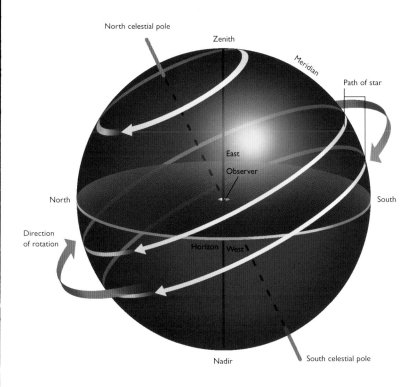

ABOVE PATHS OF THE STARS
Observers at mid-latitudes in the Northern Hemisphere will see most of the stars rise in the east, arc through the sky, and then set in the west. The stars near the north celestial pole will describe complete circles in the sky and will always be visible.

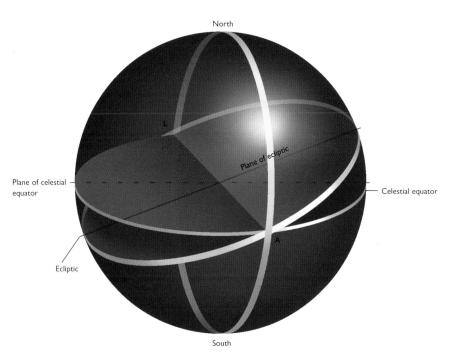

ABOVE CROSSING POINTS
The Sun crosses the celestial equator at two points on its annual path (ecliptic) around the celestial sphere. It crosses once at the vernal, or Spring equinox (on about March 21) at a point called the First Point in Aries. It crosses again six months later at the Fall equinox (on about September 23) at a point called the First Point in Libra.

star rises above the horizon on successive days, you'll find that it rises four minutes earlier each day.

In other words, the celestial sphere rotates once around Earth in 23 hours 56 minutes. Put another way, Earth rotates once relative to the stars in 23 hours 56 minutes. So this is Earth's true period of rotation in space and forms the basis of what is known as "sidereal time", or time relative to the stars.

Astronomers use sidereal time when they are observing. Then, the stars rise, culminate, and set at the same (sidereal) time. In other words, they are always in the same position in the sky at the same (sidereal) time.

CELESTIAL COORDINATES

We pinpoint the position of a star on the celestial sphere by a kind of map reference, akin to the method of latitude and longitude geographers use to pinpoint the position of a place on Earth. Celestial latitude is called declination (δ); celestial longitude, right ascension (R.A.).

Declination is the angular distance of a star north or south of the celestial equator, measured in degrees. Right ascension is the angular distance of the star along the celestial equator from the First Point of Aries. It is measured in sidereal hours and minutes.

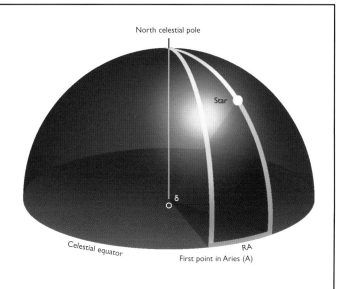

Changing Skies

The night sky is always changing. During the night the constellations move bodily across the sky, like a celestial merry-go-round. Over the course of the year, some constellations disappear from the sky and new ones take their place.

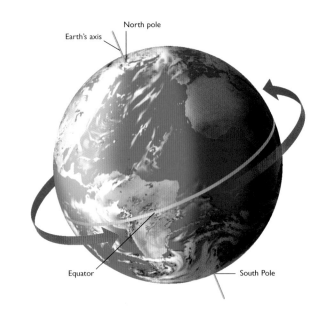

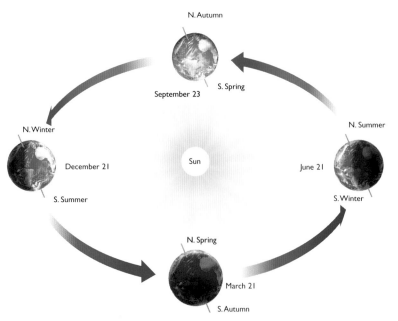

The Whirling Heavens

Remain stargazing for any length of time, and you notice that the stars move across the sky. They do not move relative to one another within the constellations, but as a whole. The stars wheel overhead because the celestial sphere rotates around Earth (see page 52). It rotates from east to west, making the stars rise above the eastern horizon and set below the western horizon.

But of course, there is actually no star-studded celestial sphere rotating around a fixed Earth once a day. It is Earth that is moving relative to the stars. It spins around once a day on its axis, an imaginary line passing through the North and South Poles. And it spins around in the opposite direction from the celestial sphere—from west to east.

It is because of the rotation of Earth that we must specify on a star map a particular time of night when the stars will appear in the sky in the positions shown on the map. At other times of the night, the stars will appear in different positions. For the star maps that appear in this and the next chapter, we show how the skies will look at about 11:00 p.m. local time.

Season by Season

When you look at the heavens over a period of months, you notice that some constellations are always present but others come and go. This happens because of Earth's other movement in space—it travels in orbit around the Sun once a year.

Again, from our Earthbound perspective, it is the Sun that appears to move. It appears to travel once a year around the celestial sphere along the ecliptic against the background of stars.

At any time, we will not be able to see the constellations that lie in the same direction as the Sun, because the Sun's glaring light will blot them out—it will be daylight. As the Sun proceeds around the celestial sphere during the year, different constellations will be blotted out in turn, while others return to dark skies.

The star maps that follow show how the skies change between winter and summer in both the Northern and Southern Hemispheres. Using these maps will help you to familiarize yourself with the constellations and get your celestial bearings.

ABOVE LEFT IN A SPIN
Earth spins on its axis once a day—the reason why the stars appear to wheel overhead during the night. Relative to the Sun, Earth spins around once in 24 hours. Relative to the stars, it spins around once in four minutes or less. We show the axis tilted to reflect the fact that it is tilted in relation to the plane of Earth's orbit around the Sun.

LEFT THE FOUR SEASONS
Significant points in the orbit of Earth around the Sun. The tilt of Earth's axis always stays the same.

OPPOSITE ROUND IN CIRCLES
During the night, the stars appear to arc through the sky, leaving circular trails in long-exposure photographs. This photograph shows in the foreground telescopes at the Roque de los Muchachos Observatory on La Palma in the Canary Islands. Being in the Northern Hemisphere, the stars appear to circle around Polaris, close to the north celestial pole.

Winter Skies in the Northern Hemisphere

Winter brings some of the most splendid constellations to northern skies, brilliant in the cool, clear atmosphere. Most magnificent of all is Orion, which dominates the southern aspect of the sky. Also more evident in Winter is the Milky Way, which bisects the heavens.

Brilliant Beacons

Orion represents a mighty hunter, facing up to the charging bull, represented by Taurus. Both constellations are outstanding. The two brightest of Orion's stars, the 1st-magnitude Betelgeuse and Rigel, provide a nice contrast. The one is noticeably reddish, the other pure white.

The three bright stars that form Orion's "belt" lead us to two other memorable stars. To the northwest they point to Aldebaran, which marks the red eye of the bull. To the southwest they point to Sirius in Canis Major, the brightest star in the whole heavens.

And there are many pleasures still to come: the pearly white arch of the Milky Way that separates Orion and Canis Major from Gemini and Canis Minor; Gemini's twin 1st-magnitude stars Castor and Pollux; and Canis Minor's star, Procyon. Overhead, near the zenith, is another beacon star, Capella in Auriga. This completes a truly remarkable concentration, for the Northern Hemisphere at least, of eight 1st-magnitude stars.

Northern Views

The northern aspect of the winter sky shows Pegasus in the west, Leo in the east, and Cygnus—flying along the Milky Way—in between. Unmistakable too is the W-shape of Cassiopeia, also embedded in the Milky Way. Like the Big Dipper, the most recognizable pattern in Ursa Major, it never sets.

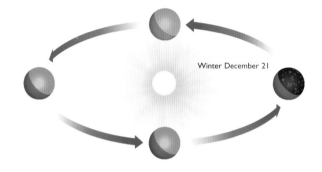

Winter December 21

BELOW LOOKING WEST
This hemisphere map shows the main constellations you would see looking west at about 11:00 p.m. local time on December 21. They may appear slightly higher or slightly lower in the sky, depending on your exact latitude.

The reddish Aldebaran, marking the eye of Taurus, the Bull, and the Pleiades are prominent high in the sky. The Square of Pegasus is unmistakable closer to the horizon.

ABOVE USING THIS MAP
This map represents the dome of the heavens in the Northern Hemisphere at about 11:00 p.m. local time on the date of the Winter Solstice, December 21.

Because the stars rise four minutes earlier each night, you will see a similar view of the heavens on December 14 at about 11:30 p.m. and on December 28 at about 10:30 p.m.

The center of the map is your zenith. The edge of your map is your horizon. You may be able to see a little more or a little less north or south than the map shows, because your horizon depends on your exact latitude.

OPPOSITE LOCATING THE CONSTELLATIONS
Looking south: to locate the constellations in the southern part of the sky, face south. The Sun will have set on your right. Hold the map in front of you with SOUTH at the bottom. The lower half of the map now represents the part of the sky in front of you.

Looking north: to locate the constellations in the northern part of the sky, face north. The Sun will have set on your left. Turn the map upside-down so that NORTH is at the bottom. The lower half of the map now represents the part of the sky in front of you.

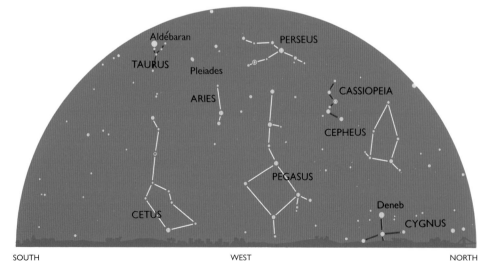

NORTHERN HEMISPHERE, WINTER

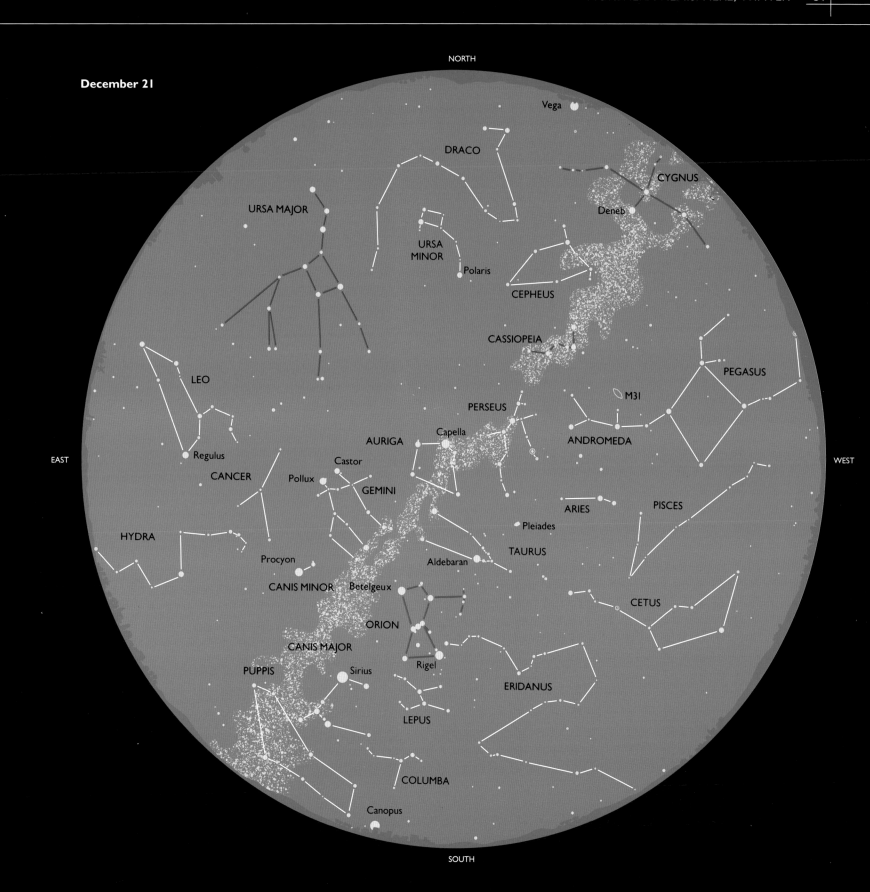

December 21

Summer Skies in the Northern Hemisphere

The constellations of Summer are nowhere near as brilliant as those of Winter. Also, evening stargazing is hindered due to the skies never growing completely dark. In Summer, northern astronomers can best glimpse a few of the gems of far southern skies, such as Scorpius.

Summer Triangle

High in the sky at the time of the Summer Solstice on June 21 are a trio of 1st-magnitude stars—Deneb in Cygnus, Vega in Lyra, and Altair in Aquila. These three beacon stars form the celebrated Summer Triangle. They span a rich region of the Milky Way.

Deneb, at the tail end of the swan (Cygnus), appears the dimmest of the Summer trio. But in reality it is the brightest—it looks dimmer because it lies much farther away than Vega and Altair.

Deneb shines high in the northeast. Andromeda and Pegasus have appeared over the eastern horizon and will become prominent in Fall skies. Leo is close to setting in the west.

Southern Delights

Rising just above the southern horizon in Summer are two of the far southern constellations that northern astronomers drool over. They are Sagittarius and Scorpius, whose pulsating red supergiant star Antares, "Rival of Mars," marks the scorpion's heart.

Antares forms one corner of a more widely separated "Summer Triangle" with Spica in Virgo, at about the same elevation, and Arcturus in Boötes farther north.

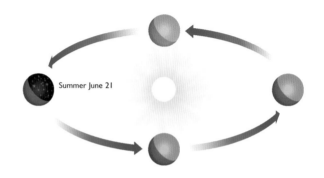

BELOW LOOKING WEST
This hemisphere map shows the main constellations you would see looking west at about 11:00 p.m. local time on June 21. They may appear slightly higher or slightly lower in the sky, depending on your exact latitude.

Arcturus in mid-skies forms one corner of a prominent triangle with brilliant white Spica and the reddish-hued Antares.

ABOVE USING THIS MAP
This map represents the dome of the heavens in the Northern Hemisphere at about 11:00 p.m. local time on the date of the Summer Solstice, June 21.

Since the stars rise four minutes earlier each night, you will see a similar view of the heavens on June 14 at about 11:30 p.m. and on June 28 at 10:30 p.m.

The center of the map is your zenith. The edge of your map is your horizon. You may be able to see a little more or a little less north or south than the map shows, because your horizon depends on your exact latitude.

OPPOSITE LOCATING THE CONSTELLATIONS
Looking south: to locate the constellations in the southern part of the sky, face south. The Sun will have set on your right. Hold the map in front of you with SOUTH at the bottom. The lower half of the map now represents the part of the sky in front of you.

Looking north: to locate the constellations in the northern part of the sky, face north. The Sun will have set on your left. Turn the map upside-down so that NORTH is at the bottom. The lower half of the map now represents the part of the sky in front of you.

NORTHERN HEMISPHERE, SUMMER 59

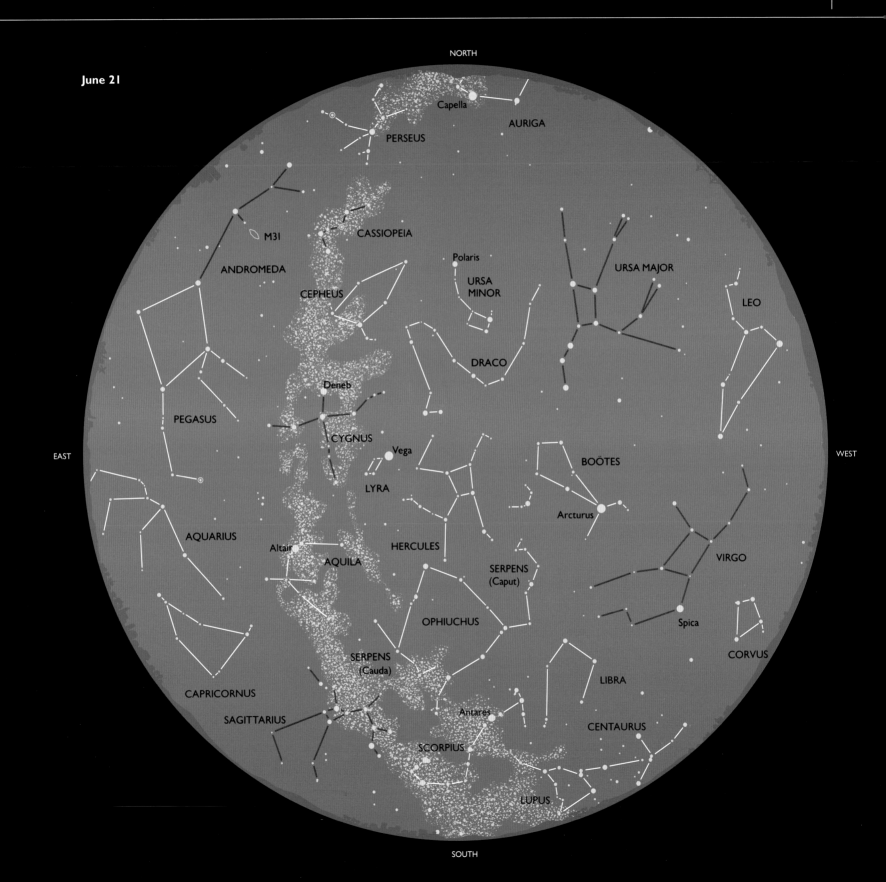

Winter Skies in the Southern Hemisphere

In June, while the Northern Hemisphere enjoys the long, warm days of Summer, the Southern Hemisphere experiences the short, cooler days of Winter. One great delight in southern Winter skies is Scorpius, instantly recognizable as a scorpion poised ready to strike.

Astronomical Delights

Scorpius is truly a magnificent constellation, set in one of the richest parts of the Milky Way. Brilliant reddish Antares, which marks the heart of the Scorpion, lies near the zenith at this time.

The adjacent constellation Sagittarius is also embedded in the Milky Way and is outstanding too. It is awash with stunning star clusters, colorful nebulae, and brilliant star fields. The whole region looks amazing when swept with binoculars.

The northern aspect of the sky shows, south of Sagittarius, three bright stars—Altair in Aquila, Vega in Lyra, and Deneb in Cygnus. They form a kind of Winter Triangle, the equivalent of the Summer Triangle northern astronomers see in their skies.

Just two bright stars are visible in the northwest, Spica in Virgo and Arcturus in Boötes. Close by is the half-circle of stars named the Northern Crown (Corona Borealis).

Looking South

The southern aspect of the southern Winter sky reveals Crux, the Southern Cross, descending. Its bright pointers, Alpha and Beta Centauri, locate it well, preventing confusion with the False Cross nearby formed by a quartet of stars in Vela and Carina. Alpha Centauri has the distinction of being the nearest bright star to us—it is only 4.3 light-years away!

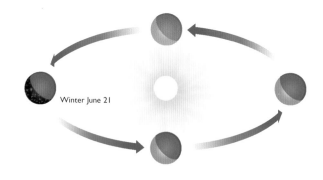

ABOVE USING THIS MAP
This map represents the dome of the heavens in the Southern Hemisphere at about 11:00 p.m. local time on June 21.

Because the stars rise four minutes earlier each night, you will see a similar view of the heavens on June 14 at about 11:30 p.m. and on June 28 at 10:30 p.m.

The center of the map is your zenith. The edge of your map is your horizon. You may be able to see a little more or a little less north or south than the map shows, because your horizon depends on your exact latitude.

BELOW LOOKING WEST
This hemisphere map shows the main constellations you would see looking west at about 11:00 p.m. local time on June 21. They may appear slightly higher or slightly lower in the sky, depending on your exact latitude.

Crux and its pointers, Alpha and Beta Centauri, are bright in mid-skies, while Spica and Arcturus shine prominently toward the north. Antares is near the zenith.

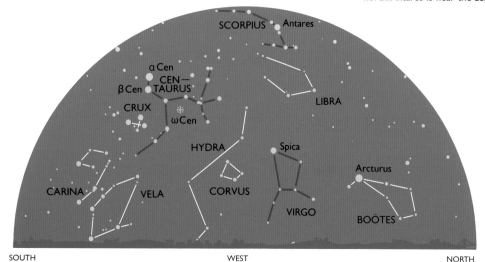

OPPOSITE LOCATING THE CONSTELLATIONS
Looking south: to locate the constellations in the southern part of the sky, face south. The Sun will have set on your right. Hold the map in front of you with SOUTH at the bottom. The lower half of the map now represents the part of the sky in front of you.

Looking north: to locate the constellations in the northern part of the sky, face north. The Sun will have set on your left. Turn the map upside-down so that NORTH is at the bottom. The lower half of the map now represents the part of the sky in front of you.

SOUTHERN HEMISPHERE, WINTER 61

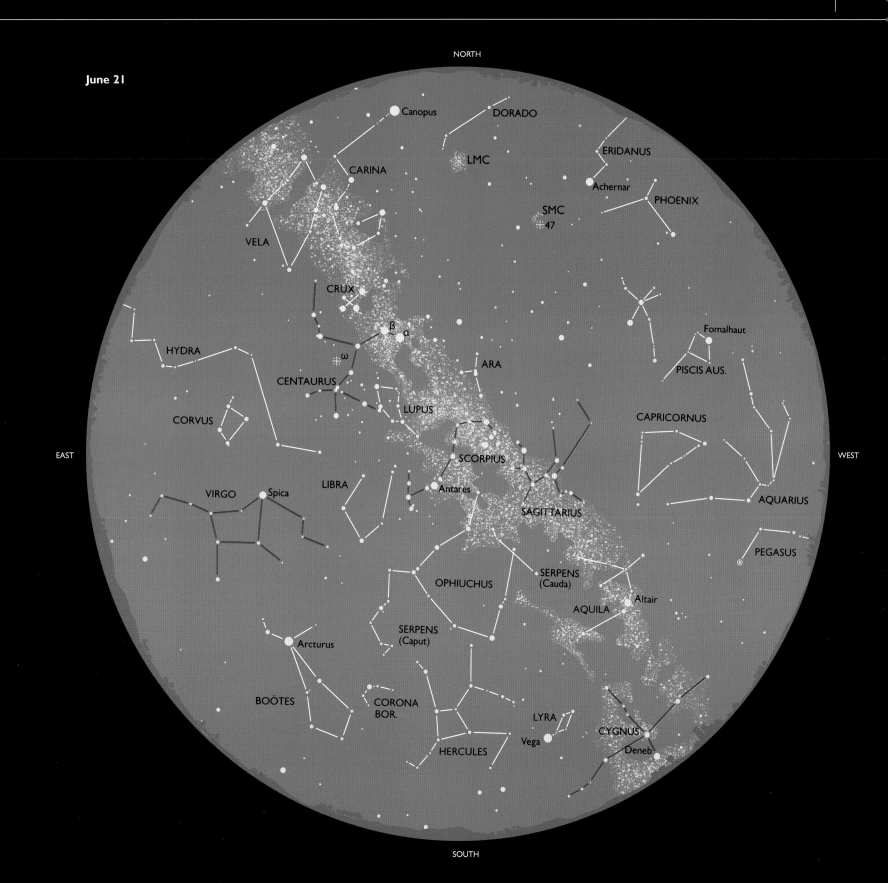

Summer Skies in the Southern Hemisphere

The brilliant constellations of northern Winter also appear, of course, in the southern Summer skies, but they are not quite as spectacular because of the lighter nights. Orion again dominates the sky, while brilliant Sirius now has a rival, the second brightest star, Canopus.

Summer Stunners

Southern astronomers see Orion upside-down compared with the view of their colleagues in the north. Rigel appears upper left of the main star pattern, while Betelgeuse appears bottom right.

As ever, Orion is a good signpost to other stars. Following a line through the three stars of Orion's "belt" leads higher up to the brightest star in the heavens, Sirius in Canis Major, and lower down to the reddish Aldebaran in Taurus. Continuing the same line through Aldebaran leads to the Pleiades, or Seven Sisters, star cluster.

Near the horizon beneath Aldebaran lies Capella in Auriga. Farther east are the twin bright stars of Gemini—Castor and Pollux—and, higher in the sky, Procyon in Canis Minor.

Looking South

The southern aspect of southern skies is spectacular. The second brightest star in the heavens, Canopus in Carina, appears high up, as does Achernar in Eridanus farther west.

Crux and Alpha and Beta Centauri, are prominent, as usual, and appear quite close to the horizon. Two other highlights are the misty patches of the Magellanic Clouds—the Large Magellanic Cloud (LMC) in Dorado and the Small Magellanic Cloud (SMC) in Tucana.

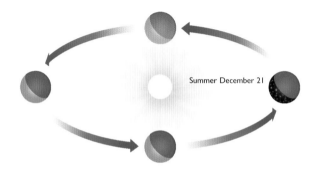

Summer December 21

BELOW LOOKING WEST
This hemisphere map shows the main constellations you would see looking west at about 11:00 p.m. local time on December 21. They may appear slightly higher or slightly lower in the sky, depending on your exact latitude.

The faint and sprawling constellations Eridanus and Cetus occupy much of the sky. Achernar, Fomalhaut, and Aldebaran are scattered highlights.

ABOVE USING THIS MAP
This map represents the dome of the heavens in the Northern Hemisphere at about 11:00 p.m. local time on December 21.

Because the stars rise four minutes earlier each night, you will see a similar view of the heavens on December 14 at about 11:30 p.m. and on December 28 at about 10:30 p.m.

The center of the map is your zenith. The edge of your map is your horizon. You may be able to see a little more or a little less north or south than the map shows, because your horizon depends on your exact latitude.

OPPOSITE LOCATING THE CONSTELLATIONS
Looking south: to locate the constellations in the southern part of the sky, face south. The Sun will have set on your right. Hold the map in front of you with SOUTH at the bottom. The lower half of the map now represents the part of the sky in front of you.

Looking north: to locate the constellations in the northern part of the sky, face north. The Sun will have set on your left. Turn the map upside-down so that NORTH is at the bottom. The lower half of the map now represents the part of the sky in front of you.

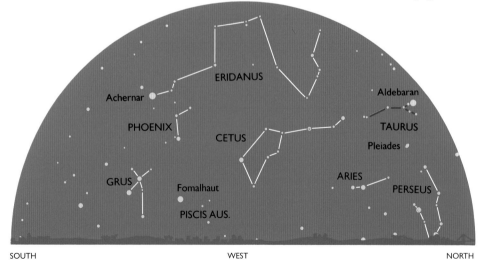

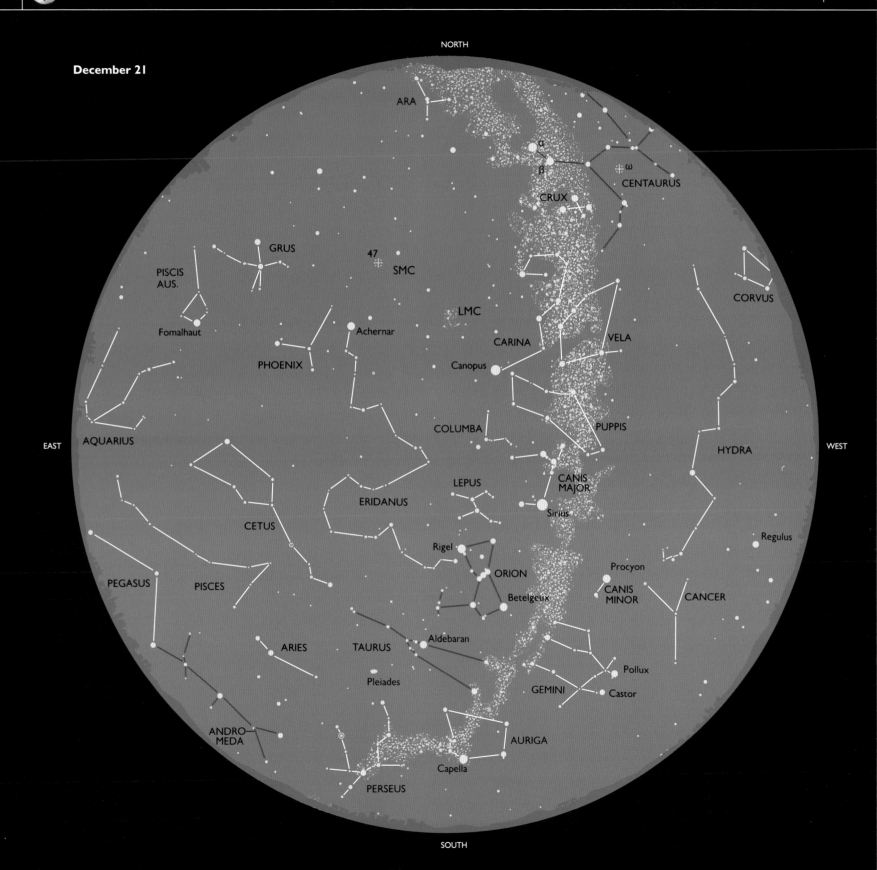

4 The skies month by month

Month by month, the constellations we see in the night sky change perceptibly as Earth travels on its annual journey around the Sun. Each constellation takes it in turn to climb to its highest point in the heavens, which happens on the meridian—the imaginary north-south line in the night sky.

In this chapter, we show the principal constellations visible each month near the meridian in a series of monthly maps that represent segments of the celestial sphere. We show them as they appear at about 11:00 p.m. local time (a time needs to be specified because the heavens spin around during the night due to Earth's rotation). For many of the months, we feature in more detail one of the constellations that is of particular interest.

Not all observers throughout the world will be able to see every constellation visible in the monthly maps: it depends on the observer's latitude, or distance north or south of the Equator. For example, an observer in Canada will not be able to see the far southern constellations that observers in Australia, for example, will see.

Two maps precede the monthly maps. They are centered on the northern and southern celestial poles, and represent the "top" and "bottom" of the celestial sphere. To far northern and far southern observers, many of these constellations are circumpolar, which means that they can be seen every night of the year.

LEFT CARINA SPECTACULAR
This spectacular vista of countless stars and swathes of glowing gas is located in the far southern constellation Carina (the Keel). The gas cloud surrounds the unstable giant star Eta Carinae, which periodically erupts and spews vast masses of gas and dust into space. Known as the Eta Carinae Nebula, the cloud is split by dark dust lanes.

FAR LEFT **LOVELY LAGOON**
This beautiful gas cloud is one of many gems in the constellation Sagittarius. It is designated NGC 6523 and M8—number 8 in Charles Messier's list.

CENTER **CLASSIC CLUSTERS**
Visible to the naked eye in the constellation Perseus is this pair of open star clusters. Also called the Double Cluster and the Sword Handle, the two clusters are variously designated h (left) and Chi Persei, and NGC 869 and 884.

ABOVE **RIVETING RING**
This colorful celestial smoke ring is the famous Ring Nebula (M57) in the constellation Lyra. It is a planetary nebula, an expanding cloud of gas puffed out by a dying star at its center.

Introducing the Monthly Maps

On the ordinary maps we use on Earth, we locate places according to a grid reference of latitude and longitude. The latitude is a measure (in degrees) of the distance a place is north or south of the Equator. The longitude is a measure (in degrees) of the distance the place is east or west of a fixed point (a line through Greenwich, England, called the Greenwich Meridian).

We use a similar grid system of celestial latitude and celestial longitude to pinpoint the positions of stars on the celestial sphere. Celestial latitude is the distance a star is north or south of the celestial equator. It is termed the star's declination (using the Greek symbol for delta, (δ), and is measured in degrees.

Celestial longitude is the distance a star is along the celestial equator from a fixed point (the First Point of Aries). It is termed the star's right ascension (RA), and is measured in hours of sidereal, or astronomical, "star time."

STAR BRIGHT

The maps feature stars down to 5th-magnitude brightness, just above the naked-eye limit of visibility of 6th-magnitude. Fainter stars are shown if they are of interest.

The brighter stars in the constellations are labeled with letters of the Greek alphabet (see opposite), according to a system introduced by the German astronomer Johann Bayer in the 1600s. In the Bayer system, the brightest star in the constellation is designated Alpha (α), the next brightest Beta (β), the next Gamma (γ), and so on. On the maps that follow, often only the stars featured in the text are labeled, making identification easier.

We also know many bright stars by proper names. For example, Alpha Canis Majoris, the brightest star in the heavens, is also known as Sirius. The proper names for many other familiar stars appear on the maps.

Most stars shine steadily in the night sky, but others vary in brightness. On the maps, some variables are identified by the Bayer letter. But others are identified differently. The first variable discovered in a constellation is designated R, the next S, and so on. After Z, the variables become RR, RS, and so on. On the maps, two symbols are used for variables: ⊙ for variables that remain visible to the naked eye at minimum brightness, and O for variables that fall below naked-eye visibility.

DEEP-SKY OBJECTS

By deep-sky objects, we mean nebulae, open star clusters, globular clusters, and galaxies. Those discussed in the text are marked on the maps with their own symbols, shown in the Key. They are usually identified either by an M number, such as M57, or simply by a number, such as 4662.

The M number is the Messier number. It is the number that was assigned to that particular object in a catalog of more than 100 nebulae and clusters by the French

LOCATION

Location Maps

For this new edition of *The Star Guide*, we have introduced an additional useful feature for the monthly maps. With a small location box (such as the one pictured above), we have pinpointed the positions of the pictures which accompany the text in the monthly maps.

INTRODUCING THE MONTHLY MAPS

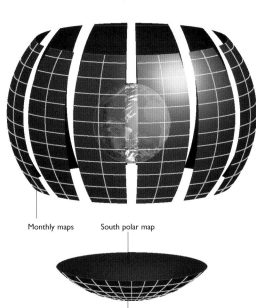

North celestial pole

North polar map

Monthly maps — South polar map

South celestial pole

This diagram shows how the monthly star maps that follow relate to the celestial sphere. The celestial sphere is shown "exploded" into 12 segments—one for each month of the year.

Each segment covers two hours of Right Ascension. It forms that central part of one of the monthly maps (see right). The stars, of course, appear on the inner surface of the segment. The north polar and south polar star maps are derived from the northern and southern "caps" of the celestial sphere, centered on the celestial poles.

RIGHT
This is a smaller version of one of the monthly maps (October), with annotations to point out salient features.

Ⓐ This lighter area shows the position of the Milky Way.
Ⓑ Symbol of a planetary nebula. The number is its NGC number.
Ⓒ Symbol for an open cluster. The number is its NGC number.
Ⓓ Symbol for a galaxy. The M number is its number in Messier's list.
Ⓔ This dashed line shows the ecliptic, the apparent path of the Sun across the celestial sphere.
Ⓕ The line of zero declination, marking the celestial equator.
Ⓖ The Greek letter Beta is the Bayer letter, which indicates that this star is the second brightest in the constellation.
Ⓗ Star name. Some of the best-known stars are labeled with their name as well as their Bayer letter.
Ⓘ Name of the constellation in Latin.
Ⓙ Scale of declination, or celestial latitude. The minus sign applies in the Southern Hemisphere.
Ⓚ Scale of right ascension, or celestial longitude.

BELOW RIGHT
Also included on each spread with the monthly star map are two "sky views" like this. One shows what an observer in mid-latitudes in the Northern Hemisphere would see looking south at about 11pm local time in the first week of the month. The other shows what an observer in the Southern Hemisphere would see looking north at the same time.

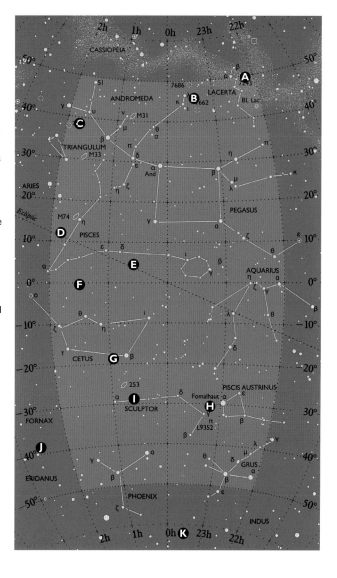

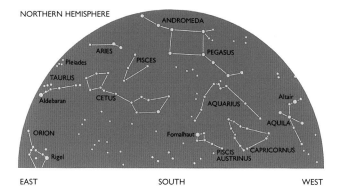

astronomer Charles Messier in the late-18th century. Messier was an ardent comet-hunter (nicknamed the "comet-ferret") and compiled the list to pinpoint misty objects that might otherwise be confused with new comets.

A number by itself next to a symbol of a deep-sky object denotes the NGC number of that object. This is the number assigned to it in the New General Catalog of Nebulae and Clusters of Stars compiled by the Danish astronomer Johann Dreyer, first published in 1888.

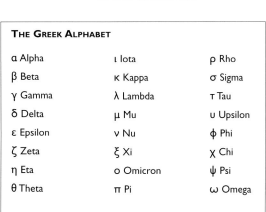

THE GREEK ALPHABET

α Alpha	ι Iota	ρ Rho
β Beta	κ Kappa	σ Sigma
γ Gamma	λ Lambda	τ Tau
δ Delta	μ Mu	υ Upsilon
ε Epsilon	ν Nu	φ Phi
ζ Zeta	ξ Xi	χ Chi
η Eta	ο Omicron	ψ Psi
θ Theta	π Pi	ω Omega

North Polar Stars

Looking north in the Northern Hemisphere, observers can see some of the constellations circling overhead throughout the year. They are the circumpolar constellations: Ursa Major, Ursa Minor, Draco, Cepheus, and Cassiopeia. The north polar region is centered on Polaris, the Pole Star, which scarcely changes its position in the sky.

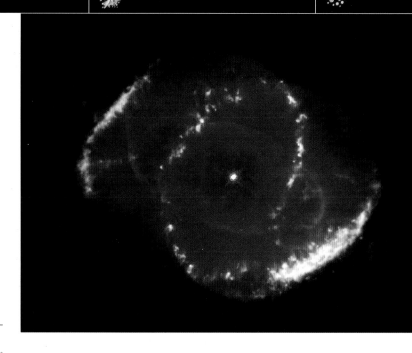

Cassiopeia

This key constellation is featured overleaf See page 70

Cepheus

King Cepheus, whose wife was the vain Queen Cassiopeia, and whose daughter was Andromeda—offered as sacrifice to Cetus

Although Cepheus is a relatively faint constellation, with no stars brighter than 2nd-magnitude, it can be readily identified by its house-like outline. Alternatively, you can find it by using Alpha (α) and Beta (β) Cassiopeiae as pointers.

Alpha (α) Cephei, with a magnitude of 2.4, is interesting because it will become the pole star in about 7,500 years time after precession has changed the geometry of the heavens.

Delta (δ) Cephei's claim to fame is that it is the prototype for the class of variable stars known as Cepheid variables, noted for their regular-as-clockwork change in brightness. Delta Cephei varies with absolute precision between magnitudes 3.5 and 4.4 in 5.4 days. The star is also a binary star, with a bluish 6th-magnitude companion, which can be visually separated via a small telescope.

However, the most visually interesting star in the constellation is Mu (μ). One of the reddest stars in the sky, it was named the Garnet Star by William Herschel. Its hue is noticeable to the eye, but is best seen with binoculars. Mu is a pulsating red giant, which varies in magnitude between 3.5 and 5 over a period of about two years.

Draco, The Dragon

The multi-headed monster Ladon guarded the golden apples in the beautiful garden of the Hesperides. It was slain by Hercules

This sprawling constellation—the eighth largest in the heavens—winds itself, serpent-like, around the celestial north pole.

Alpha (α), also called Thuban, is not in fact the brightest star; it is more than a magnitude less than the brightest, 2nd-magnitude Nu (ν). Alpha is interesting because it was the pole star about 4,500 years ago. A passage in the Great Pyramid of Giza in Egypt, constructed around this time, was aligned with this star.

Nu, in the Dragon's head, is one of several fine double stars in the constellation. The equally bright 5th-magnitude components can easily be separated in binoculars. They are both white.

Enclosed by the neck of the Dragon, roughly midway between Zeta (ζ) and Delta (δ), is one of the finest planetary nebulae in the heavens. It is NGC 6543, or the Cat's Eye Nebula. In small telescopes it presents a noticeably bluish disk, although the central white star is difficult to make out. The Hubble Space Telescope, however, captures the extraordinary beauty of the Cat's Eye.

Ursa Major, The Great Bear

The key constellation featured in March See page 84

Ursa Minor, The Little Bear

Represents Ida, one of the two nymphs (the other was Adrasteia) who nursed the infant Zeus, who became king of the gods

Also known as the Little Dipper, this constellation is notable because of its brightest star, Alpha (α), which is named Polaris. This star is currently the Pole Star, located less than one degree from the north celestial pole. It appears to remain fixed in the heavens, while all the other stars whirl around it during the night.

Polaris is a 2nd-magnitude star, best found by using two stars in the Big Dipper (Merak and Dubhe) as pointers (see page 84). It is a Cepheid variable star, although its slight variations in brightness are difficult to spot. Small telescopes will spot that it has a faint (8th-magnitude) bluish companion. Eta (η) and Gamma (γ) are also optical doubles.

Beta (β), in the bowl of the dipper, has about the same brightness as Polaris, and may be confused with it. But Beta is noticeably yellower.

LOCATION

ABOVE THE CAT'S EYE
The planetary nebula NGC 6543 in Draco is one of the most stunning objects in the heavens. Well named the Cat's Eye Nebula, it has the most intricate structure, created through repeated ejections of matter from the central white dwarf star.

NORTH POLAR STARS 69

Cassiopeia, distinctive northern circumpolar constellation

Cassiopeia is an unmistakable constellation, with its brightest stars forming a distinctive W-shaped pattern. It lies far north, on the other side of Polaris, the Pole Star, from Ursa Major. Like that constellation, Cassiopeia is circumpolar from Canada and the northern United States and northern Europe.

ABOVE **THE VAIN QUEEN**
Queen Cassiopeia was the wife of King Cepheus and mother of Andromeda. She boasted that she was fairer than the Nereids, sea nymphs noted for their beauty. The sea god Poseidon punished Cassiopeia for her vanity by ordering that Andromeda be sacrificed to the monster Cetus.

THE "W" STARS

The constellation's five brightest stars define its W-shape. Alpha (α), Beta (β), and Gamma (γ) are almost identical in brightness (about magnitude 2.2).

Alpha is the southernmost star of the W, also called Shedir. It is noticeably orange compared with Beta. Eta (η), the closest bright star to Alpha, is a fine double star, with golden yellow and purplish components.

Extending a line through Delta (δ) and Epsilon (ε) an equal distance brings you to Iota (ι). Small telescopes will reveal that this is a multiple star, with white, yellow, and blue components.

A VARIETY OF VARIABLES

Continuing the same line through Iota locates RZ Cassiopeiae, an eclipsing binary variable star, and one of the few to be within the range of binoculars. Usually a steady 6th-magnitude, it dims to nearly 8th-magnitude every 29 hours when its brightest component is eclipsed.

At the opposite end of the constellation lies R Cassiopeiae, a Mira-type, long-period variable, which varies between magnitudes 6 and 13 over a period of 431 days.

North of R, and not too far from Beta, lies an arc of three stars Tau (τ), Rho (ρ), and Sigma (σ), all roughly of the 5th-magnitude. Rho is an interesting variable of unknown type, which fluctuates between magnitude 4 and 6 quite unpredictably.

FINE CLUSTERS

Much of Cassiopeia lies within the Milky Way, in which star clusters and nebulae abound. It is a rich region for sweeping with binoculars.

Rich star fields lie around Epsilon, Delta, and Gamma together with a number of clusters. Between Epsilon and Delta lie NGC 663, NGC 654, and M103. South of Delta, next to Phi (φ), lies NGC 457, which is the brightest, with a magnitude of about 6.5. It is a compact group of more than 100 stars. Binoculars show it well, and small telescopes reveal its individual stars.

M52 is another rich cluster found near the edge of the constellation by extending a line through Alpha and Beta. It is a cluster of hot stars, like the more familiar Pleiades in Taurus. Small telescopes show it well.

LOCATION

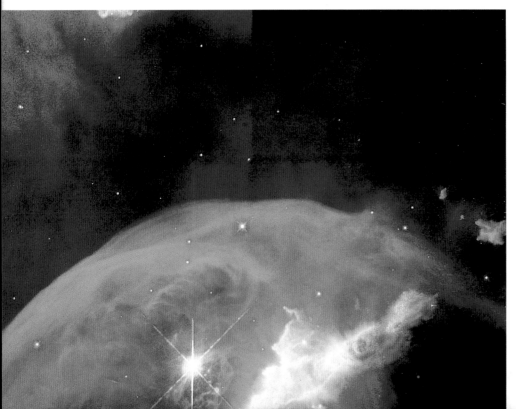

LEFT **A BILLOWING BUBBLE**
This expanding shell of glowing gas in Cassiopeia is NGC 7635, aptly named the Bubble Nebula. Located just south of the cluster M52, it is a difficult subject for small telescopes. But here, the Hubble Space Telescope reveals its breathtaking beauty.

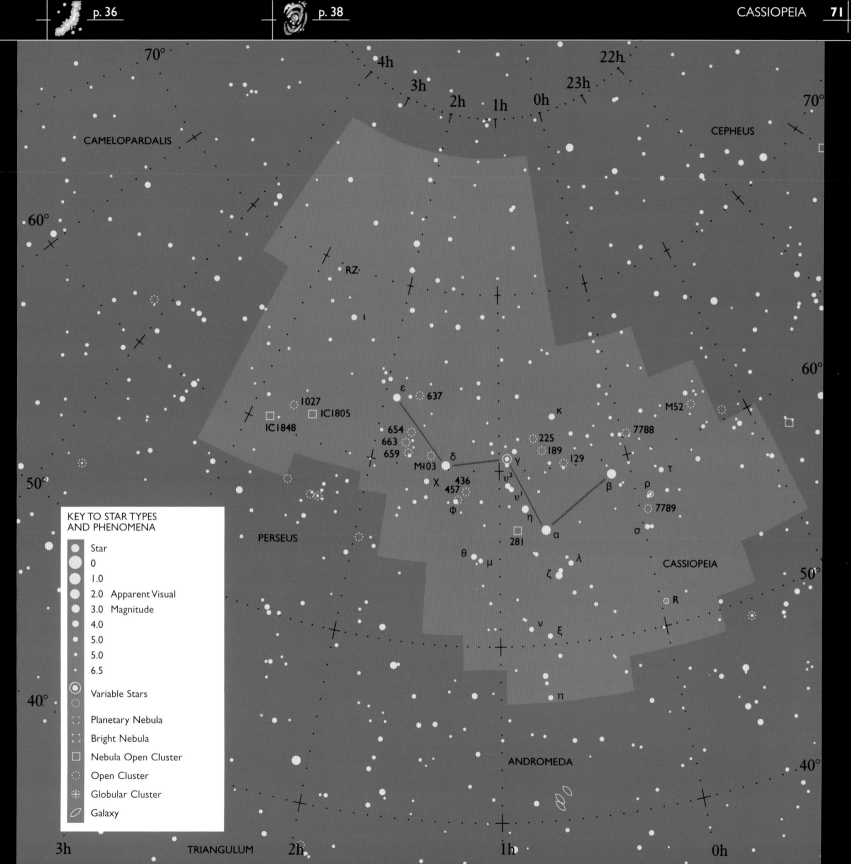

South Polar Stars

In contrast with northern skies, southern skies have no convenient pole star. Indeed, the central polar region is remarkably empty. But the long axis of the unmistakable Crux, the Southern Cross, points roughly in the direction of the south celestial pole. Other outstanding features of far southern skies include the nearest galaxies, the famed Magellanic Clouds.

ARA, THE ALTAR

Either the altar at which the gods swore their oaths or the one on which the Centaur (Centaurus) was going to sacrifice the wolf (Lupus)

One of the smaller constellations, Ara lies mostly in the Milky Way. One of its highlights is NGC 6397, a fine globular cluster just visible to the naked eye and spectacular through binoculars or a small telescope. It is one of the closest globulars we know, at around 8,000 light-years away.

CARINA, THE KEEL

Of the ship (Argo Navis), in which Jason and the Argonauts set sail in search of the fabled Golden Fleece

Carina is a spectacular constellation on the edge of the Milky Way, boasting the second brightest star in the heavens, Canopus (magnitude -0.7).

The constellation extends nearly to Crux, the Southern Cross. Its stars Iota (ι) and Epsilon (ε), along with two stars in the adjacent constellation Vela (Delta (δ) and Kappa), form a False Cross. This is sometimes confused with the true Cross, Crux.

Eta (η) Carinae is one of the most fascinating stars in the constellation. This massive, unstable star is subject to periodic eruptions. It is embedded in a glorious nebula, the Eta Carinae Nebula (NGC 3372), just visible to the naked eye and delightful via binoculars and small telescopes (see picture page 64).

Close to Epsilon (ε) is a fine open cluster, NGC 2516, while another (IC 2602) surrounds bluish-white Theta (θ). Both are visible to the naked eye.

CENTAURUS, THE CENTAUR

*This key constellation is featured overleaf
See page 74*

CRUX, THE SOUTHERN CROSS

*This key constellation is featured overleaf
See page 74*

DORADO, THE SWORDFISH

A modern constellation, introduced in the 16th century

Dorado is noted not for its stars, but because it is home to a large misty patch we call the Large Magellanic Cloud (LMC). This "Cloud" is not a nebula in our own Galaxy, but a galaxy in its own right.

The LMC is in fact the nearest galaxy to our own, and is one of the few clearly visible to the naked eye. Around 30,000 light-years across, it lies about 170,000 light-years away.

The brightest feature of the LMC, visible to the naked eye, is the Tarantula Nebula (NGC 2070), also called 30 Doradus. It was in this nebula that the brightest supernova seen for centuries erupted in 1987.

TUCANA, THE TOUCAN

Another 16th-century creation

Although its stars are insignificant, Tucana is notable because it includes the Small Magellanic Cloud (SMC) and the glorious globular cluster 47 Tucanae.

The SMC is another neighboring galaxy, quite close in the sky to the Large Magellanic Cloud in Dorado. 47 Tucanae looks like a 4th-magnitude star to the eye, but binoculars or small telescopes reveal that it is a dense mass of close-packed stars.

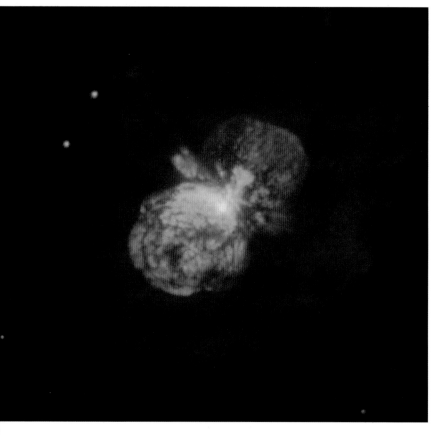

ABOVE FLARING UP
The giant star Eta Carinae erupted violently in the 1830s, spewing vast clouds of gas and dust into space. Here, the Hubble Space Telescope reveals the expanding cloud of matter that originated from the eruption.

LOCATION

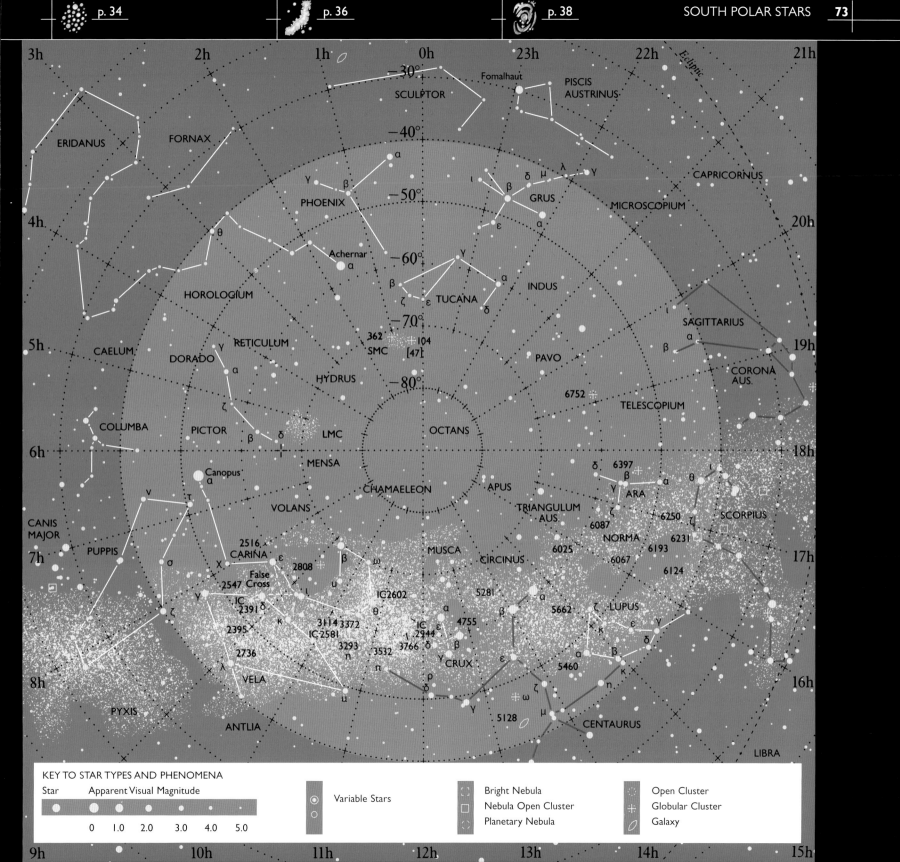

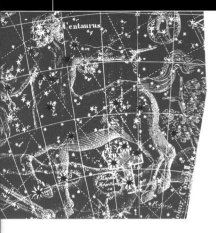

Centaurus, the Centaur
Crux, the Southern Cross

Centaurus is a huge constellation, the ninth largest in the heavens. It envelops Crux, which is the smallest. Whereas Centaurus has mythological connections, Crux does not, being a 16th-century creation. Both constellations occupy a rich part of the heavens, dipping into one of the brightest regions of the Milky Way.

ABOVE THE LEARNED CENTAUR
Half-man, half-horse, centaurs were generally wild creatures. But the centaur this constellation depicts was different. Named Chiron, he was wise and learned, and achieved fame as a great teacher, particularly of hunting, music, and medicine.

STUPENDOUS CENTAURUS

Centaurus is an outstanding constellation, boasting the nearest star to the Sun, the third brightest star in the sky, the finest globular cluster, and one of the strongest radio sources in the heavens. Northern Hemisphere observers lament that they can only catch a glimpse of such a sumptuous constellation.

The two brightest stars in Centaurus are the distinctive pair Alpha (α) and Beta (β) Centauri. They act as a useful signpost in southern skies, pointing to Crux, the Southern Cross.

The brightest of the two is Alpha, with a magnitude of -0.3. It is outshone in the heavens only by Sirius and Canopus. Alpha Centauri, sometimes called Rigil Kent, is a double star, separated in small telescopes into yellow components. It is the nearest bright star, located just 4.3 light-years away, less than half the distance to Sirius.

Larger telescopes reveal a faint 11th-magnitude dwarf star in orbit around Alpha, known as Proxima ("nearest") Centauri because it is the nearest star to the Sun, at about 4.2 light-years away.

Between Alpha and the noticeably whiter and fainter Beta is the variable star R. It is a Mira-type variable, which is an easy binocular object at maximum brightness (5th-magnitude) but fades over a period of 547 days to 11th-magnitude, well beyond binocular range.

LOCATION

LEFT A VERITABLE JEWEL
It was William Herschel's son John who named this beautiful open cluster in Crux the Jewel Box. Located around the star Kappa Crucis, its stars flash all the colors of the rainbow, like sparkling jewels.

CLUSTERS AND GALAXIES

Following a line through Beta and Epsilon (ε) to an equal distance beyond brings you to what appears to be a 4th-magnitude star, designated Omega (ω) Centauri. But it is not a star. Binoculars reveal that it is a concentrated globe-shaped mass of stars—in other words a globular cluster. It is in fact the brightest of all globular clusters, although the other outstanding globular cluster 47 Tucanae runs a close second. Both are brighter than 4th-magnitude.

Among other clusters worth investigating via binoculars or a small telescope are another globular, NGC 5286, just north of Epsilon, and the open cluster NGC 3766, near Lambda (λ).

SMALL BUT BEAUTIFUL

Tiny Crux is the most distinctive southern constellation, and is circumpolar from New Zealand and much of Australia. Of the four main stars that make up the cross, Alpha (α), or Acrux, is the brightest at magnitude 0.8, somewhat brighter than Beta (β) and Gamma (γ). Gamma contrasts visibly with the other two stars because it is a warm orange rather than white. Gamma is also an optical double star, like Beta and Mu (μ).

Near Beta is the constellation's finest gem. It is an open cluster of colored stars centered on Kappa (κ) Crucis. Its popular name is the Jewel Box. At its edge lies a "hole" in the Milky Way, in reality a dark nebula named the Coal Sack.

LOCATION

ABOVE RICHEST GLOBULAR
The globular cluster Omega Centauri is one of the finest sights in Centaurus. It is the brightest globular in the whole heavens, brighter than 4th-magnitude. Even small telescopes will reveal its stars.

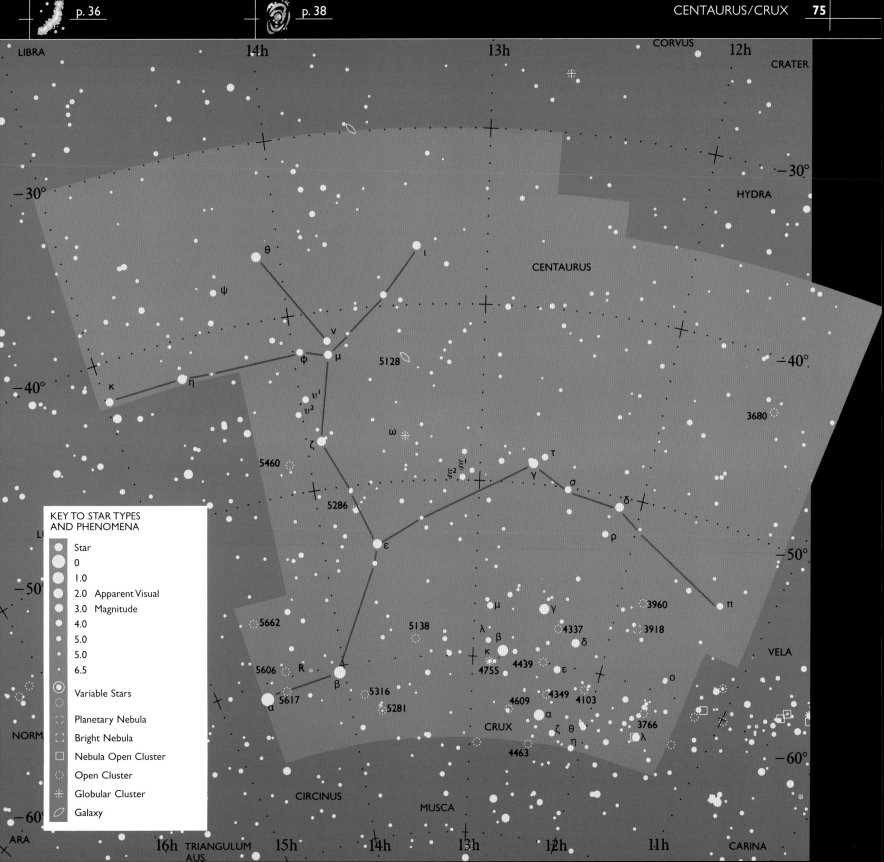

January Skies

The skies this month are among the most dazzling of the year. They are dominated by the familiar constellations of Gemini, Taurus, Orion, and Canis Major. And they are lit up by a surfeit of bright stars that shine like beacons—Capella, Castor, Pollux, Aldebaran, Procyon, Betelgeuse, Rigel, and Sirius.

AURIGA, THE CHARIOTEER

The inventor of the four-horse chariot, Erichthonius, whose father was the Greek god of fire Hephaestus (the Roman god Vulcan)

This conspicuous kite-shaped constellation lies astride the Milky Way. It can quickly be located by its lead star Alpha (α), or Capella.

Capella is the sixth brightest star in the heavens and is noticeably yellowish in hue. It is actually a spectroscopic binary, consisting of two giant stars orbiting close together.

South of Capella is a triangle of fainter stars, known as the Heidi, or the Kids (Kids meaning young goats, not children!). Of the three, Epsilon (ε) and Zeta (ζ) are both eclipsing binaries. In the Epsilon system, a dark companion passes in front of a brilliant supergiant star every 27 years, when the brightness halves from about magnitude 2.8 to 3.8.

Auriga's deep-sky objects—set in the Milky Way—include three fine, open clusters M38, M36, and M37, which lie more or less in a line just south of Theta (θ). All three are easy binocular objects of about 6th-magnitude.

CANIS MAJOR, THE GREAT DOG

One of the dogs that Orion used for hunting, with the hare (Lepus), at his feet

This constellation is famous because it boasts the brightest star in the heavens, Sirius, also known as the Dog Star. With an apparent magnitude of -1.45, it shines like a beacon in January skies.

However, Sirius is not truly bright: it appears so only because it is relatively close to us, at a distance of under nine light-years—a mere stone's throw in the vast expanses of the Universe.

Sirius is a binary star with a close companion that orbits every 50 years or so. This companion of Sirius, or Sirius B, is also known as the Pup. Although it is 8th-magnitude, it is difficult to spot because it tends to be obscured by Sirius's glare.

Sirius B is not a normal star like Sirius (A), but a white dwarf, the remnant of a dying Sun-like star. It was the first white dwarf to be discovered—by U.S. astronomer Alvan Clark in 1862.

Two fine open clusters are found in the constellation. Easiest to spot is 4th-magnitude M41, located due south of Sirius. It consists of around 100 colorful stars, easily picked up with binoculars. NGC 2362 is equally bright, but more difficult to make out because it lies in the Milky Way. Its several dozen stars cluster around Tau (τ).

COLUMBA, THE DOVE

Introduced in the 16th-century; the dove sent by Noah to seek dry land

Relatively small and with no really bright stars, Columba is easy to spot because it lies in a featureless region of the heavens. A suitable object for small telescopes is the globular cluster NGC 1851.

ABOVE STAR MAGNITUDES Using the symbols on the left, this map plots the star magnitudes and other visible objects in January skies.

KEY TO STAR TYPES AND PHENOMENA

Star
- 0
- 1.0
- 2.0 Apparent Visual
- 3.0 Magnitude
- 4.0
- 5.0
- 5.0
- 6.5

- Variable Stars
- Planetary Nebula
- Bright Nebula
- Nebula Open Cluster
- Open Cluster
- Globular Cluster
- Galaxy

JANUARY SKIES

LEPUS, THE HARE

The fleet-footed hare pursued across the sky by one of Orion's dogs (Canis Major)

Easily located immediately south of Orion, Lepus contains much of interest. Gamma (γ) is a lovely double star, well seen in binoculars, with gold and white components. South of Beta (β) is the globular cluster M79, suitable for small telescopes.

But the most interesting object in Lepus is the star R Leporis, found by extending a line from Alpha (α) through Mu (μ). It is one of the reddest stars we know and is often called the Crimson Star. This star is a Mira variable that changes in brightness between 5th- and 10th-magnitude over a period of about 14 months.

ORION

*The key constellation featured in January
See page 78*

ABOVE THE GOAT STAR
Capella in Auriga is the brightest star in this image. It is the third brightest star in northern skies and the sixth brightest overall, with a magnitude of 0.1. Capella is sometimes known as the Goat Star.

LOCATION

BELOW NORTHERN HEMISPHERE
View of the night sky from latitudes 40 to 50 degrees north, looking south at about 11:00 p.m. local time on about January 7.

BELOW RIGHT SOUTHERN HEMISPHERE
View of the night sky from latitudes 35 to 40 degrees south, looking north at about 11:00 p.m. local time on about January 7.

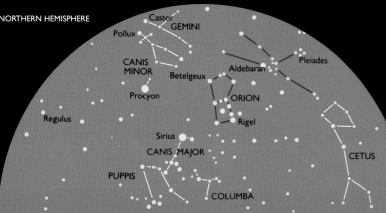

Orion, the most magnificent constellation in the heavens

Straddling the celestial equator, Orion is equally familiar to observers in both the Northern and Southern Hemispheres. Its pattern of bright stars is distinctive and can readily be visualized as the figure of a mighty hunter, with a club raised ready to strike in one hand, and a shield in the other.

ABOVE THE MIGHTY HUNTER
Orion was famed as a hunter. The son of the sea god Poseidon, he features in many mythological tales. In one, he fell in love with seven sisters, the Pleiades, and he persued them, as he does still in the heavens.

LOCATION

RIGHT THE GREAT NEBULA
The Orion Nebula (M42) is the brightest and best-known nebula in the heavens. Located some 1,500 light-years away, it measures around 15 light-years across. The small nebula shown here above M42 is designated M43, but it is part of the same nebulosity as the big nebula.

BETELGEUSE AND RIGEL

Two bright supergiant stars dominate the constellation—Alpha (α), named Betelgeuse, and Beta (β), named Rigel. Brilliant, white Rigel is actually the brighter of the two, of magnitude 0.1. It is an intensely hot star, with the energy output of 50,000 Suns. Noticeably reddish, Betelgeuse varies in brightness over a period of about six years. It can reach magnitude 0.5, but can dip below magnitude 1.5 at times. Betelgeuse is truly supergigantic, with a diameter of around 250 million miles (400 million km). If it were where the Sun is in our Solar System, it would stretch nearly as far as the planet Mars.

GREAT NEBULAS

In the figure of Orion, three bright stars make up his sword belt: Zeta (ζ), Epsilon (ε), and Delta (δ). A bright misty patch marks the position of the Sword Handle in the sky.

Binoculars and small telescopes reveal that this patch is a huge glowing nebula, aptly named the Great Nebula in Orion. But long-exposure photographs are needed to bring out the stunning beauty of this object.

The whole constellation is embedded in vast gas clouds, and the Orion Nebula belongs to a particularly dense region known as the southern molecular cloud. Another dense region, around Zeta, is known as the northern molecular belt.

Just south of Zeta is the glowing nebula IC 434, which has a mass of dark gas silhouetted against it in the shape of a horse's head. This dark mass, also named Barnard 33, is better known as the Horsehead Nebula.

MULTIPLE CHOICE

The gas that forms the Orion Nebula is set glowing by the powerful radiation from young hot stars embedded within it. One of the stars, Theta (θ) Orionis, when viewed in a telescope, can be seen to be a quadruple star, named the Trapezium because of the arrangement of its four components.

Sigma (σ), just south of Zeta, is also a multiple star, and small telescopes can make out its red, white, and blue components. Zeta itself is a double star, as is Delta. So are the bright Iota (ι), south of the Orion Nebula, and Lambda (λ), the star that marks Orion's "eye."

Another highlight of the constellation is the variable W, located near the southern end of the arc of Pi (π) stars that form Orion's shield. W is a long-period variable of the Mira type and varies between 6th- and 8th-magnitude over a period of 212 days. Its brightening and dimming can readily be perceived with binoculars.

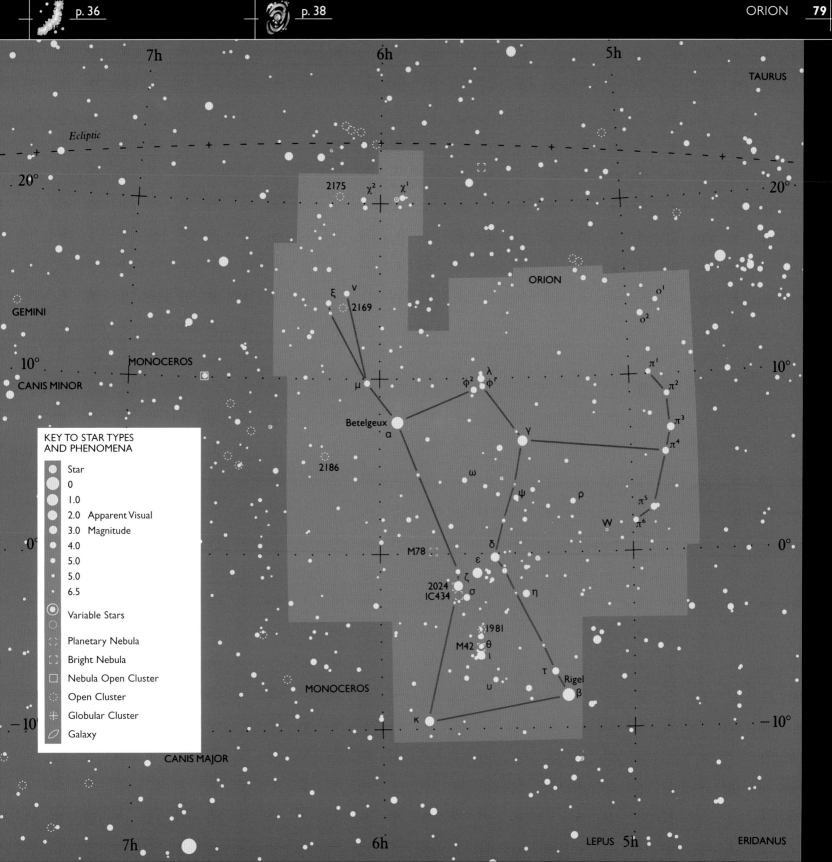

THE MONTHLY MAPS p. 20 p. 24 p. 28

February Skies

The January spectacle of brilliant stars slips west, to be replaced by skies with fainter constellations, such as Cancer and Hydra. But Castor and Pollux in Gemini, Procyon in Canis Minor, Sirius in Canis Major, and Betelgeuse and Rigel in Orion still delight. Leo climbs steadily in the east, heralding a change in the seasons.

CANCER, THE CRAB

The crab that Zeus's wife Hera sent to kill Hercules, which he trod on and killed

Cancer is one of the constellations of the zodiac—in astrology, the fourth sign of the zodiac.

It is small and faint but still contains much of interest. Iota (ι), for example, is a lovely double star, with gold and pale blue components. But Cancer's highlight is the beautiful open cluster M44, which is called Praesepe, or the Beehive.

Of 3rd-magnitude, Praesepe can easily be seen with the naked eye close to a line joining Gamma (γ) and Delta (δ). It probably contains as many as 300 individual stars. M67, near Alpha (α), is another cluster of hundreds of stars, visible in binoculars.

Cancer also features two lovely double stars Iota (ι) and Zeta (ζ). Both have yellow and blue components, visually separated via small telescopes.

CANIS MINOR, THE LITTLE DOG

The smallest of the two dogs Orion used when he went hunting

Canis Minor is a small and generally uninteresting constellation, notable only for its lead star Alpha (α), called Procyon. At magnitude 0.4, Procyon is the eighth brightest star in the heavens. Like the other "dog star," Sirius, it is a double star with a white dwarf companion. This tiny star orbits Procyon about once every 40 years.

GEMINI, THE TWINS

The twins Castor and Pollux, the sons of Leda, queen of Sparta, and her seducer, Zeus

Gemini is one of the constellations of the zodiac—in astrology, the third sign of the zodiac. It is well named because it has twin 1st-magnitude stars, Castor and Pollux. Pollux is the brighter of the two and is more colorful, being a rich orange-yellow.

Castor is rather more interesting because it is a multiple-star system. Small telescopes show it to be a binary star, and a spectroscope reveals that each component is also double. Another eclipsing binary belongs to the system too, making six stars in all.

For binoculars, a good deep-sky object in Gemini is the open cluster M35. Another for small telescopes is NGC 2392, a fine planetary nebula, called the Eskimo.

MONOCEROS, THE UNICORN

A modern constellation, introduced in the 17th century

Monoceros is a faint constellation that straddles a rich region of the Milky Way. Close to Epsilon (ε) there is a fine open cluster, NGC 2244. Larger telescopes reveal that the cluster is surrounded by a lovely flower-like nebula, well named the Rosette.

Just to the north of NGC 2244 is another open cluster, NGC 2264. It too is surrounded by a striking nebula, known as the Cone Nebula.

KEY TO STAR TYPES AND PHENOMENA

Star
0
1.0
2.0 Apparent Visual
3.0 Magnitude
4.0
5.0
5.0
6.5

Variable Stars
Planetary Nebula
Bright Nebula
Nebula Open Cluster
Open Cluster
Globular Cluster
Galaxy

ABOVE STAR MAGNITUDES Using the symbols on the left, this map plots the star magnitudes and other visible objects in February skies.

PUPPIS, THE POOP

Of the ship Argo Navis, in which Jason and the Argonauts sailed when they went in search of the Golden Fleece

Much of this far southern constellation lies in the Milky Way and abounds in rich star fields and clusters.

The lead star Zeta (ζ) is 2nd-magnitude and is one of the hottest stars we know, with an electric blue color. L2, near Sigma (σ), is of interest because it is a long-period variable whose brightness varies between 3rd- and 6th-magnitude over about five months.

Two open clusters, NGC 2451 and 2477, located between Zeta and Pi (π), are worth viewing with binoculars. The latter is particularly dense, looking almost like a globular cluster.

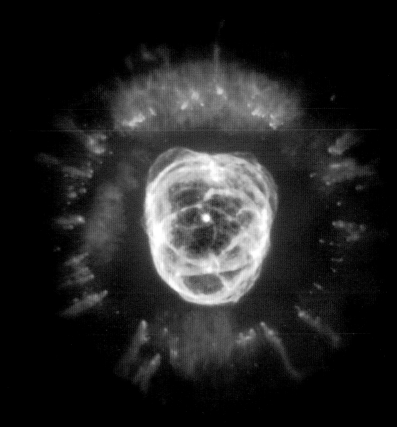

ABOVE THE ESKIMO
This planetary nebula in Gemini is NGC 2392, captured here spectacularly by the Hubble Space Telescope. It is named the Eskimo because it looks like a face surrounded by an eskimo's parka.

LOCATION

BELOW NORTHERN HEMISPHERE
View of the night sky from latitudes 40 to 50 degrees north, looking south at about 11:00 p.m. local time on about February 7.

BELOW RIGHT SOUTHERN HEMISPHERE
View of the night sky from latitudes 35 to 40 degrees south, looking north at about 11:00 p.m. local time on about February 7.

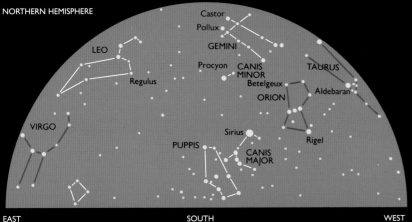

NORTHERN HEMISPHERE

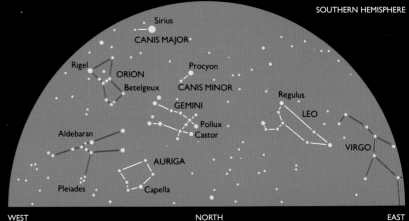

SOUTHERN HEMISPHERE

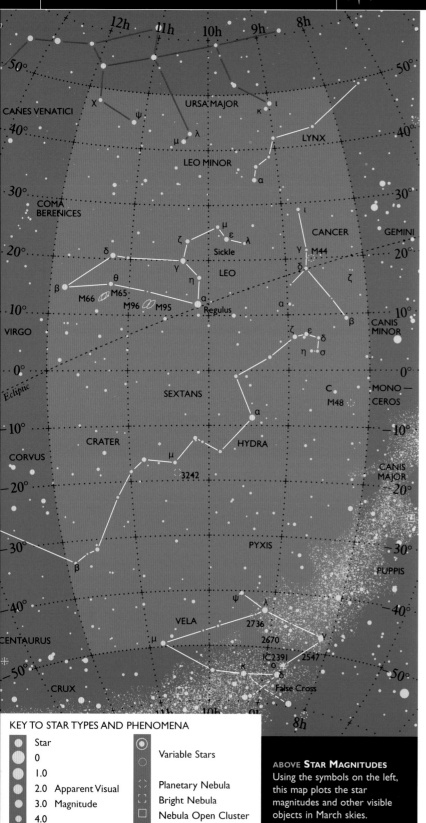

March Skies

While Orion is slipping toward the western horizon, Leo is climbing into the mid-sky. Its presence there reminds us that spring has arrived in the Northern Hemisphere, and fall in the Southern. Otherwise, the skies are relatively bare, being occupied primarily by the sprawling Hydra, largest of all the constellations.

HYDRA, THE WATER SNAKE (HEAD)

A multi-headed serpent with poisonous breath, which Hercules killed on the second of his 12 Labors; he dipped his arrows in the Hydra's blood, making them deadly

Hydra extends more than a quarter of the way around the celestial sphere. But it is not a conspicuous constellation: the only really bright star is 2nd-magnitude Alpha (α), named Alphard, which means "the solitary one." It is also known as Cor Hydrae, or the Hydra's Heart. A giant red star, it has a distinctly orange hue.

The stars in the Hydra's head make an attractive group when viewed via binoculars. Of them, Epsilon (ε) is a binary star, whose components need a telescope to separate.

M48, to the south of the group, is a fine open cluster, just visible to the naked eye and very well seen with binoculars and small telescopes.

(For Hydra's tail region, see page 86.)

LEO, THE LION

The Nemean lion, which Hercules slew on the first of his 12 Labors. Its skin was invincible, so Hercules had to strangle it.

The star pattern of this constellation of the zodiac looks passably like the figure of a crouching lion. The lion's head and mane are defined by a pattern of stars that make the shape of a backward question mark or sickle, and so it is named the Sickle.

The first star in the handle of the Sickle is 1st-magnitude Alpha (α), or Regulus. Also known as Cor Leonis, or the Lion's Heart, it is seen to be a double star via binoculars, with an 8th-magnitude companion.

Among the other Sickle stars, Gamma (γ), Mu (μ), and Epsilon (ε) are noticeably orange. Gamma is also called Algeiba, which means the Lion's Mane. Small telescopes show it to be a lovely binary system, with golden-yellow components. Brightest of the trio of stars that define the tail end of the lion is 2nd-magnitude Beta (β), also known as Denebola.

Among the deep-sky objects visible in small telescopes in the constellation are two pairs of galaxies: M65 and M66 lie immediately south of Theta (θ); M95 and M96 lie roughly between them and Regulus. All four galaxies are spirals and of about 9th-magnitude.

LYNX, THE LYNX

A modern constellation, introduced in the 17th century

This faint northern constellation has only one fairly bright star, 3rd-magnitude Alpha (α). There are a few double and multiple stars for telescope observers. They include 12 and 38, which are both triple stars.

URSA MAJOR, THE GREAT BEAR

The key constellation featured in March See page 84

ABOVE STAR MAGNITUDES Using the symbols on the left, this map plots the star magnitudes and other visible objects in March skies.

VELA, THE SAILS

Of the ship Argo Navis, in which Jason and the Argonauts set sail to search for the Golden Fleece

The brightest stars in this far southern constellation are between 2nd- and 3rd-magnitude. Small telescopes show that the brightest, Gamma (γ), is a double. The brightest of the two components also proves to be a spectroscopic binary. One of the binary components is a Wolf-Rayet star, particularly hot and luminous and thought to be on the brink of exploding as a supernova.

Visually, the stars Kappa (κ) and Delta (δ) form a distinctive cross shape with two stars (Iota and Epsilon) in the adjacent constellation Carina. This pattern can sometimes be confused with the true Southern Cross, Crux, and is therefore called the False Cross.

Most of Vela lies in the Milky Way, and is therefore rich in star fields, clusters, and nebulae. Centered on the star Omicron (o) is the bright open cluster IC 2391, while NGC 2591 is a fainter one south of Gamma.

LOCATION

ABOVE VELA SPECTACULAR
Rich in star fields, nebulae, and clusters, the Milky Way in Vela is a stunning sight through binoculars and small telescopes.

BELOW NORTHERN HEMISPHERE
View of the night sky from latitudes 40 to 50 degrees north, looking south at about 11:00 p.m. local time on about March 7.

BELOW SOUTHERN HEMISPHERE
View of the night sky from latitudes 35 to 40 degrees south, looking north at about 11:00 p.m. local time on about March 7.

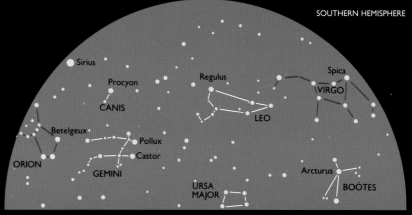

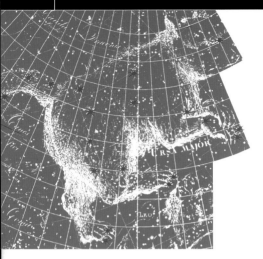

Ursa Major, the Great Bear
the signpost of northern skies

The third largest constellation, Ursa Major, is the most familiar northern constellation because it contains the conspicuous star pattern known as the Big Dipper, or in Europe, as the Plow. It is circumpolar for much of the Northern Hemisphere. A line through two of its stars, Merak and Dubhe, point to the Pole Star or North Star, Polaris, and this has made the Big Dipper a boon to navigators for centuries.

ABOVE COMELY CALLISTO
When Zeus lay with the fair huntress Callisto, his wife Hera was not amused and turned Callisto into a bear. Zeus placed her in the heavens as the Great Bear to prevent her from being killed by their son Arcas.

BELOW THE HUBBLE DEEP FIELD
In an 120-hour exposure, the Hubble Space Telescope spied thousands of distant galaxies in an apparently empty region of the heavens just north of the Big Dipper. Some lie as much as 10 billion light-years away.

LOCATION

The constellation's seven brightest stars form the distinctive pattern of the Big Dipper, so called because it resembles a ladle. In Europe, it is called the Plow because it also resembles the handle and share blade of an old horse-drawn plow.

All the Dipper's main stars have names derived from Arabic. Moving along the Dipper from the handle end, they are Alkaid (Eta), Mizar (Zeta), Alioth (Epsilon), Megrez (Delta), Phad (Gamma), Merak (Alpha), and Dubhe (Beta). Of these, the brightest stars are Dubhe and Alioth, both magnitude 1.8.

Merak and Dubhe are valuable signpost stars because if you extend a line from Merak through Dubhe, it points to the Pole Star, Polaris. That is why these two stars are called the Pointers.

SEEING STARS

The most interesting of the Dipper stars is 2nd-magnitude Mizar. It forms a classic naked-eye double with 4th-magnitude Alcor. With small telescopes, Mizar itself is seen to be double—but it is in reality a true binary system. And spectroscopes reveal that each component is a close binary too.

Ursa Major also has a number of variable stars that can be studied in binoculars. They include R, north of Dubhe, and T, which makes an equilateral triangle with Epsilon and Delta. They are both long-period variables, varying between 7th- and 13th-magnitudes. R has a period of 301 days; T a period of 265 days.

GALAXIES AND NEBULA

Ursa Major is well-endowed with galaxies, many of which are in range of small telescopes and even binoculars. The two brightest are a close pair, M81 and M82. They form one corner of a triangle with Dubhe and 23.

M81 is a classic spiral, similar to our own Galaxy. Of 7th-magnitude, it is brighter than M82, which is rather irregular in shape. M82 seems to be undergoing upheaval and is a powerful radio source.

M101, above the handle of the Dipper, is an 8th-magnitude galaxy, well seen through binoculars. Even small telescopes may reveal the open arms of this face-on spiral.

Another galaxy quite easy to find via small telescopes is the spiral M108, located close to Merak. Close by is a fainter planetary nebula, M97. Larger telescopes show it to have two dark patches within the ring-like gas cloud, which has given it the nickname, the Owl Nebula.

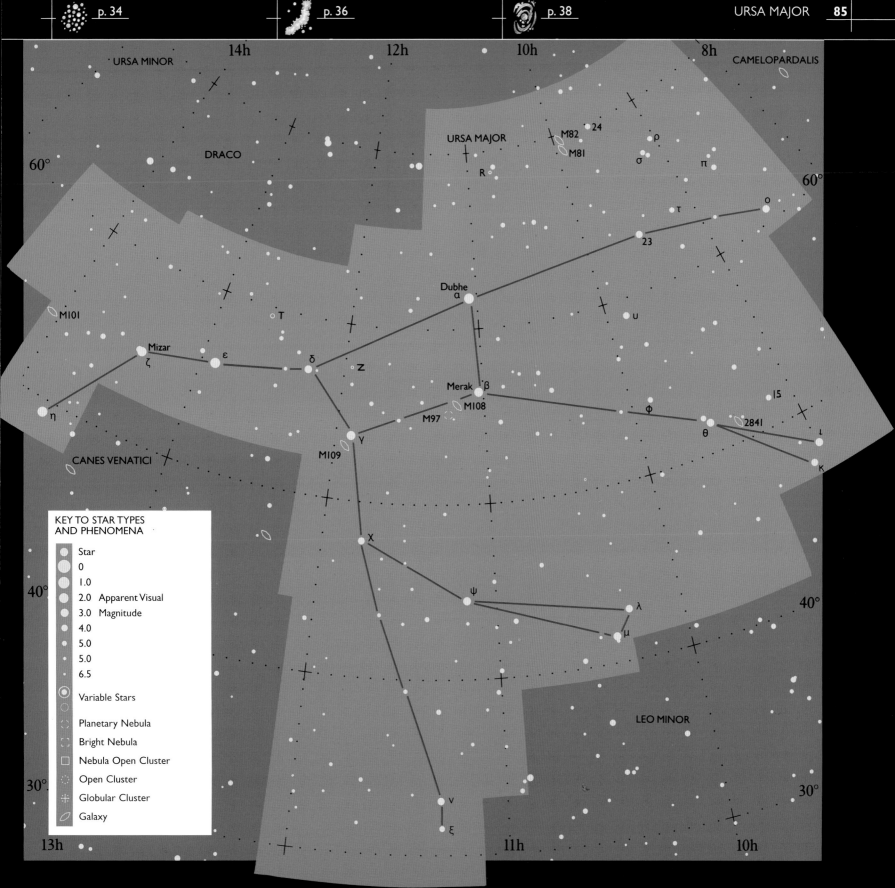

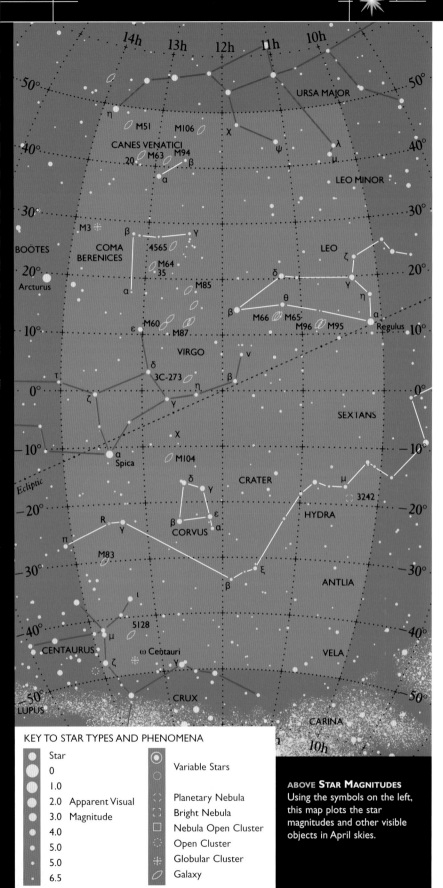

April Skies

Leo is still high in the sky this month. But most of the constellations moving in from the east are much fainter, so to the eye, the sky looks relatively bare. Two constellations dominate, Hydra and Virgo, the two biggest constellations in the heavens. Between them they boast only one 1st-magnitude star, Spica.

CANES VENATICI, THE HUNTING DOGS

A 17th-century creation, formerly part of Ursa Major, depicting two dogs held on a leash by Boötes, the Herdsman

Located under the curve of the handle of the Big Dipper, this is an inconspicuous constellation. Its lead star, the 3rd-magnitude Alpha (α), is also known as Cor Caroli, meaning Charles's Heart. It was so named to honor the English King Charles I, by the first Astronomer Royal, Edmond Halley (of Halley's Comet fame). Alpha is a double star, easily separated in small telescopes.

The real "stars" of this constellation, however, are M3 and M51. M3 is a globular cluster on the southern boundary of the constellation, just east of Beta (β) Comae Berenices, in the adjacent constellation. Of 6th-magnitude, M3 is on the limit of naked-eye visibility and easily found in binoculars and small telescopes. It vies with M13 in Hercules for the honor of being the finest globular in northern skies. Larger telescopes begin to resolve some of its closely packed stars.

M51 is the famous Whirlpool Galaxy, a beautiful face-on spiral. Located in the north of the constellation, it is probably best found by scanning southward from Eta (η) Ursa Majoris. Small telescopes will show M51 as a pair of fuzzy spots. Larger instruments identify these as the centers of two interconnected galaxies.

Historically, M51 is important because it was the first galaxy to have its spiral structure revealed, by Lord Rosse in 1845 using his huge telescope, named the Leviathan of Parsonstown.

Canes Venatici is also home to many more galaxies, including M63, M94, and M106. Like M51, they are spirals and around 9th-magnitude.

CORVUS, THE CROW

A bird sent by Apollo to fetch water in a cup (the constellation Crater), who dallied to eat figs and then blamed Hydra for the delay

The four main stars in this tiny constellation are easily seen with binoculars, all of about 3rd-magnitude. Of these, Delta (δ) is a double, resolvable in small telescopes. Several galaxies are present, but none bright enough for small instruments.

HYDRA, THE WATER SNAKE (TAIL)

For mythology, see page 82

In mid-skies this month is the tail end of Hydra. One interesting star here is R, just east of Gamma (γ). It is a Mira-type long-period variable star, easily visible when it is at its brightest (4th-magnitude). But over a period of a little over a year, it dims to 11th-magnitude, below binocular visibility.

ABOVE STAR MAGNITUDES Using the symbols on the left, this map plots the star magnitudes and other visible objects in April skies.

KEY TO STAR TYPES AND PHENOMENA

Star
- 0
- 1.0
- 2.0 Apparent Visual
- 3.0 Magnitude
- 4.0
- 5.0
- 5.0
- 6.5

- Variable Stars
- Planetary Nebula
- Bright Nebula
- Nebula Open Cluster
- Open Cluster
- Globular Cluster
- Galaxy

APRIL SKIES

LOCATION

ABOVE THE WHIRLPOOL
M51 in Canes Venatici is a fine face-on spiral with wide, open arms. It is two galaxies in one. The main one (NGC 5194) is interacting with a smaller galaxy (NGC 5195), and a bridge of stars connects them.

To the south of R is a relatively bright spiral galaxy, M83. Even larger amateur telescopes reveal that it is a lovely face-on barred-spiral galaxy.

Another impressive deep-sky object is NGC 3242, located just south of Mu (μ). Also named the Ghost of Jupiter Nebula, it is one of the brightest planetary nebulae in the heavens. Small telescopes show this 9th-magnitude planetary as a faint bluish disk.

LOCATION

ABOVE FINE GLOBULAR
The globular cluster M3 in Canes Venatici, which is just below naked-eye visibility. Binoculars and small telescopes will begin to resolve the outer fringes of this globe of some 50,000 stars.

BELOW NORTHERN HEMISPHERE
View of the night sky from latitudes 40 to 50 degrees north, looking south at about 11:00 p.m. local time on about April 7.

BELOW SOUTHERN HEMISPHERE
View of the night sky from latitudes 35 to 40 degrees south, looking north at about 11:00 p.m. local time on about April 7.

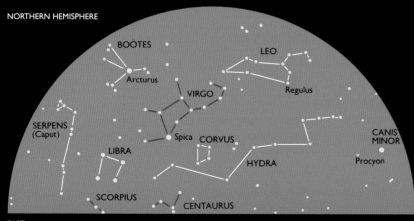

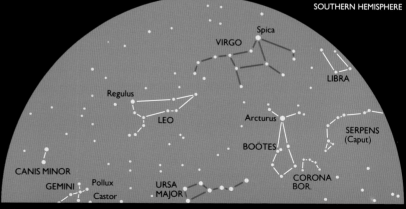

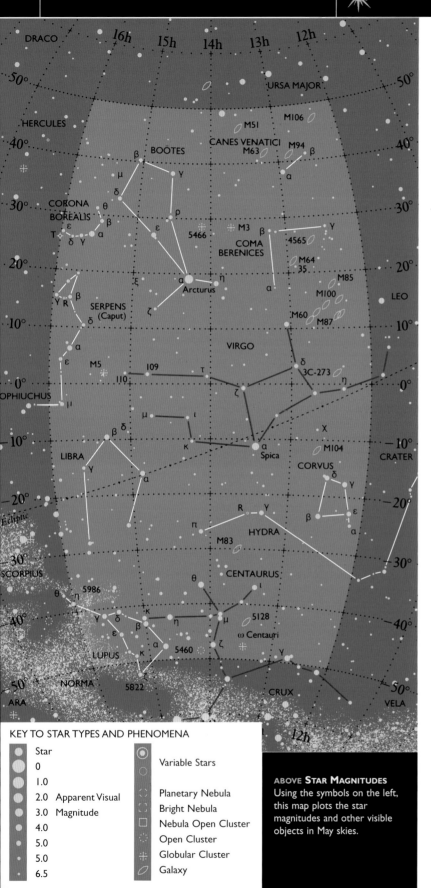

May Skies

Just two really bright stars shine in mid-sky this month: Arcturus and Spica, both close to the meridian. They provide a noticeable color contrast—Arcturus being orange, while Spica is pure white. Both are giants—one cool, the other very hot. May is the best month in the Northern Hemisphere for glimpsing some of the stars in Centaurus, low on the southern horizon.

Boötes, the Herdsman

Sometimes called the Bear Driver, chasing the Great and Little Bears (Ursa Major and Minor) across the sky. Boötes is usually identified with Arcas, the son of Zeus and one of his many conquests—the beautiful nymph Callisto.

Shaped rather like an old-fashioned kite, this constellation boasts the brightest star in the northern sky and the fourth brightest star in the whole heavens. This star is Alpha (α), named Arcturus, which means "Bear-keeper."

Arcturus has a magnitude of -0.1. It is a red giant with a pronounced orange hue and is easily located by following around the curve of stars in the handle of the Big Dipper.

There are three fine doubles in Boötes. One is Epsilon (ε), which small telescopes resolve into beautiful yellowish-orange and bluish-green components. Other doubles include Kappa (κ) and Xi (ξ); both are true binaries, with each component orbiting around the other. Mu (μ) is a wide double, easily separated into a pair of yellow stars via binoculars. Telescopes show that the fainter star of the pair is itself a binary system with yellow components.

One of the few deep-sky objects of note in Boötes is the globular cluster NGC 5466, located north of Arcturus at the same elevation as the M3 cluster in the adjacent Coma Berenices. But it is fainter and less well defined than M3.

Coma Berenices, Berenice's Hair

A creation of the 17th century; but with mythological connection. Berenice was the wife of an Egyptian king, Ptolemy. She vowed to sacrifice her hair if her husband succeeded in battle. He did—so she did

This constellation is inconspicuous to the eye, with nothing of interest in its faint stars. But binoculars give a hint of Coma's richness, showing a number of fuzzy spots that telescopes resolve into galaxies. More powerful telescopes reveal galaxies in their hundreds.

These galaxies belong to the celebrated Coma Cluster, which merges in the south with the Virgo Cluster. Altogether, this region of space contains thousands of galaxies. Most of them are beyond the range of amateur telescopes, but a few can be seen.

M64 is one, located near the star 35. Larger instruments show it as an ellipse, with a dark dust band beneath the bright center. This has given it the nickname of the Black-Eye Galaxy. Farther north, toward Gamma (γ), NGC 4565—known as the Needle Galaxy—appears needle-like because we see it edge-on.

Both the galaxies mentioned are about 9th-magnitude. So is M100, reckoned to be the brightest galaxy in the Coma Cluster, whose center is around 450 light-years away. M100 is a face-on spiral galaxy, with a bright central nucleus and wide-open spiral arms.

ABOVE STAR MAGNITUDES Using the symbols on the left, this map plots the star magnitudes and other visible objects in May skies.

KEY TO STAR TYPES AND PHENOMENA

- Star
  - 0
  - 1.0
  - 2.0 Apparent Visual
  - 3.0 Magnitude
  - 4.0
  - 5.0
  - 5.0
  - 6.5
- Variable Stars
- Planetary Nebula
- Bright Nebula
- Nebula Open Cluster
- Open Cluster
- Globular Cluster
- Galaxy

MAY SKIES 89

LIBRA, THE SCALES

Introduced by the Romans, who visualized it as the scales of justice held by the adjacent figure of Virgo

One of the faintest constellations of the zodiac, Libra was once incorporated in Scorpius, as the scorpion's claws. Alpha (α), named Zubenelgenubi ("southern claw"), is an easy double for binoculars. At magnitude 2.7 it is fractionally less bright than Beta (β), named Zubenelchemale, which has a distinct greenish hue.

Delta is also interesting because it is an eclipsing binary, like Algol in Perseus. Every 56 hours the dim star in the system eclipses the bright one, causing the overall brightness to fall from around 4.9 to 5.9 magnitude.

VIRGO, THE VIRGIN

*The key constellation featured in May
See page 90*

LOCATION

RIGHT COMA CLOSE-UP
Detailed close-up view of M100 taken by the Hubble Space Telescope. One of the multitude of galaxies in Coma Berenices, M100 is a face-on spiral with wide-open arms, visible via small telescopes.

BELOW NORTHERN HEMISPHERE
View of the night sky from latitudes 40 to 50 degrees north, looking south at about 11:00 p.m. local time on about May 7.

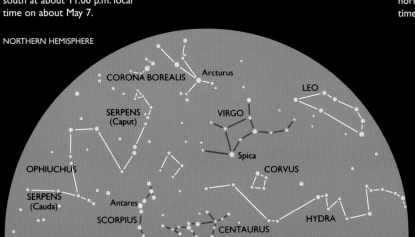

BELOW SOUTHERN HEMISPHERE
View of the night sky from latitudes 35 to 40 degrees south, looking north at about 11:00 p.m. local time on about May 7.

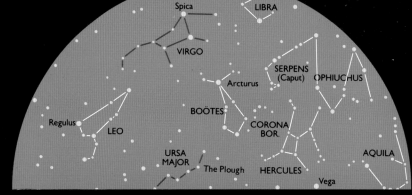

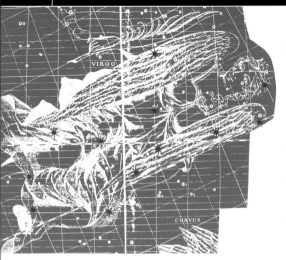

Virgo, the Virgin
home to galaxies by the thousand

Second in size only to Hydra, Virgo is a sprawling constellation of the zodiac. Apart from Spica, its stars are disappointing, and the constellation pattern is not easily followed. Virgo's astronomical interest lies in its galaxies, which cram the space from south of Spica to the northern limits of the constellation and into the adjacent Coma Berenices.

ABOVE CORN GODDESS
Virgo is depicted in the heavens with wings and holding an ear of grain. She is associated with Demeter, the goddess of agriculture and fertility. By Zeus, she had a daughter, Persephone, who was kidnapped by Zeus's brother Hades and taken to the Underworld.

EAR OF WHEAT

In the constellation figure, Spica marks the virgin's left hand, in which she holds an ear of wheat. Bluish-white in color, it is among the "top 20" brightest stars, with a magnitude of exactly one. It is a close binary system, which can be detected only with the aid of a spectroscope. Gamma (γ) is also a binary, which telescopes can resolve. Both components are yellowish and have a magnitude of 3.6.

THE VIRGO CLUSTER

The west of the constellation looks bare to the eye, but in telescopes this region appears full of galaxies. They also extend well into the neighboring constellation Coma Berenices, forming the celebrated Coma-Virgo Cluster.

At a distance of some 50 light-years, this is the closest big cluster to us. It may contain as many as 3,000 galaxies in all. A few of them may be glimpsed with binoculars and many more via small telescopes. But the majority need large telescopes to locate them and bring out their structure.

THE MESSIER GALAXIES

Some of the brighter galaxies are within range of small to medium telescopes and were cataloged by Charles Messier. They can be found just west of Epsilon (ε), near the edge of the boundary with Coma Berenices.

M59 and M60 are a close pair of elliptical galaxies, due north of 5th-magnitude Rho (ρ). M58 is a barred-spiral galaxy, M89 an elliptical, and M90 a spiral. Nearby M87 is a giant elliptical that is notable for its powerful radio output. A blue jet of matter emerges from its center, probably from a supermassive black hole that resides there.

Another interesting galaxy is M104, due south of Gamma (γ) on the border with Corvus. This galaxy can be spotted with binoculars, but a telescope is needed to see that it is a spiral that presents itself edge-on. A prominent dust lane runs through the center of the disk. Its popular name is the Sombrero Galaxy.

THE FIRST QUASAR

Located just north of Eta (η), but visible only with large telescopes, is an object called 3C-273. It shines like a star of 13th-magnitude. However, it is not of our Galaxy but lies more than 3 billion light-years away.

LOCATION

ABOVE ON THE RADIO
This near-spherical giant elliptical galaxy, M87, contains as many as 3 trillion stars. It is a radio galaxy—an active galaxy that pours out most energy at radio wavelengths rather than as light.

LOCATION

RIGHT BLACK HOLE
This donut-shaped torus of gas and dust, spied by the Hubble Space Telescope, is found at the heart of the Virgo galaxy NGC 4261. It is believed to provide the "fuel" for a supermassive black hole at its center.

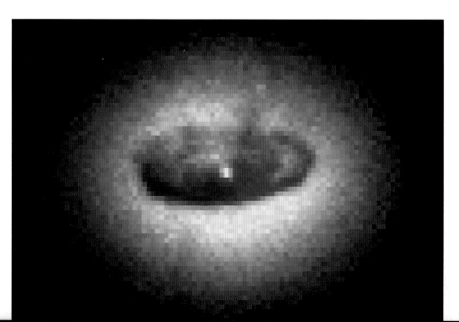

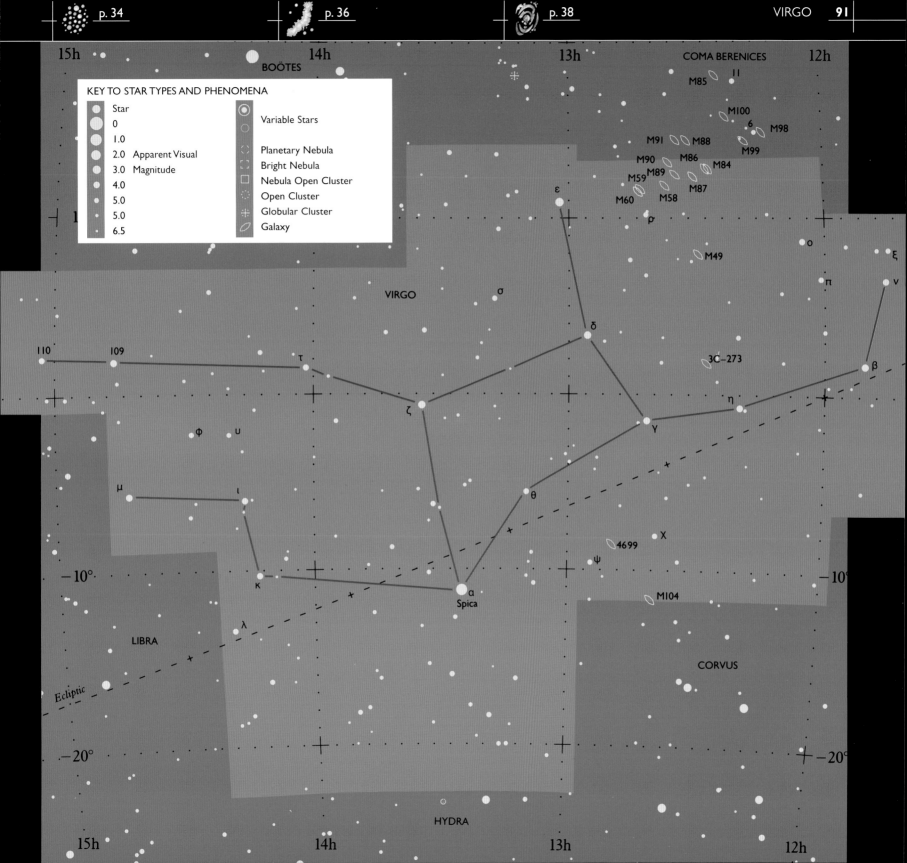

June Skies

In this month and next, observers in the Northern Hemisphere have their best views of the brilliant Scorpius, albeit low in the sky. Southern observers, on the other hand, see the constellation magnificently high overhead. Its brightest star Antares joins Arcturus, Vega, Deneb, Altair, and Spica in the night sky starscape.

Corona Borealis, the Northern Crown

The golden crown worn by the Cretan princess Ariadne, half-sister of the Minotaur, when she married the god of wine Dionysus

This tiny constellation is a neat semicircle of stars, with 2nd-magnitude Alpha (α) in the middle. It is easy to find along a line joining the two bright stars Arcturus in Boötes and Vega in Lyra, both prominent in June skies.

Nu (ν) is a double star, consisting of a pair of red giants and can be seen in binoculars. But the most interesting stars in Corona Borealis are two outstanding variables. One is R, located north of Delta (δ), easily visible in binoculars. For most of the time R, a yellow supergiant, shines at a steady 6th-magnitude, but sometimes, quite unpredictably, it dims to 12th-magnitude over a period of a few weeks. It may take months to recover its original brightness.

Another unusual variable is T, located just south of Epsilon (ε). Usually about 10th-magnitude, it can flare up in a matter of hours to naked-eye visibility.

Hercules

A son of Zeus, Hercules was one of the great Greek heroes, famous for undertaking 12 seemingly impossible tasks, or Labors.

Hercules is the fifth largest constellation, but is not particularly impressive to the eye.

Among its stars, Alpha (α) is a huge red supergiant that looks noticeably reddish in the sky. It is a variable that fluctuates between the 3rd- and 4th-magnitude. Small telescopes show it is double. Rho (ρ) is also a double.

But the outstanding feature of Hercules is M13, a fine globular cluster just visible to the naked eye. Found just south of Eta (η), M13 is the brightest globular in northern skies. M92 is another good globular, located north of Pi (π), but you need binoculars to see it. Small telescopes begin to resolve the stars in both globulars.

Lupus, the Wolf

Once considered to be any wild animal, Lupus is portrayed on star maps as the beast about to be speared by the centaur (Centaurus)

Sandwiched between Centaurus and Scorpius, Lupus inhabits a rich part of the heavens on the edge of the Milky Way. Small telescopes reveal that Pi (π), Kappa (κ), and Xi (ξ) are doubles. Mu (μ) also appears double via small instruments, but larger ones show that the brighter star of the pair is a binary system.

Being in the Milky Way, several clusters can be seen, including the open cluster NGC 5822, close to Zeta (ζ). At magnitude 6.5, it is somewhat brighter than NGC 5986, which is a small globular cluster.

ABOVE STAR MAGNITUDES Using the symbols on the left, this map plots the star magnitudes and other visible objects in June skies.

KEY TO STAR TYPES AND PHENOMENA

- Star
- 0
- 1.0
- 2.0 Apparent Visual
- 3.0 Magnitude
- 4.0
- 5.0
- 5.0
- 6.5

- Variable Stars
- Planetary Nebula
- Bright Nebula
- Nebula Open Cluster
- Open Cluster
- Globular Cluster
- Galaxy

JUNE SKIES

SCORPIUS, THE SCORPION

The key constellation featured in June
See page 94

SERPENS CAPUT, THE SERPENT'S HEAD

The head of the snake killed by Asclepius, the Greek god of medicine, which became a symbol of healing

Serpens is the only constellation split in two, by Ophiuchus. Serpens Caput is the serpent's head; Serpens Cauda, the serpent's tail (see page 97). The group of stars that form the head looks attractive via binoculars. Delta (δ) is a binary star, separated with small telescopes.

However, the gem of Serpens Cauda is the globular cluster M5, one of the finest in the heavens. Just visible to the naked eye, it is best found with binoculars south-west of Alpha (α).

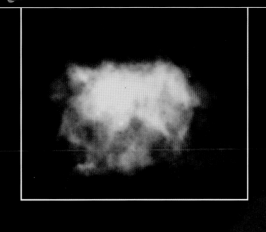

BELOW NORTHERN HEMISPHERE
View of the night sky from latitudes 40 to 50 degrees north, looking south at about 11:00 p.m. local time on about June 7.

NORTHERN HEMISPHERE

LOCATION

ABOVE THE TURTLE
The planetary nebula NGC 6210 in Hercules is well named the Turtle. The "shell", shown in red here, measures about half a light-year across. The central region has quite a complex structure.

BELOW SOUTHERN HEMISPHERE
View of the night sky from latitudes 35 to 40 degrees south, looking north at about 11:00 p.m. local time on about June 7.

SOUTHERN HEMISPHERE

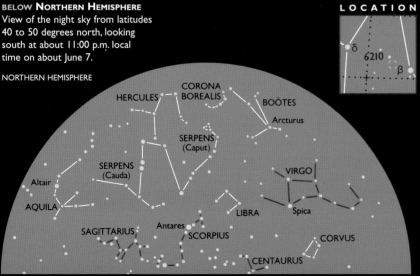

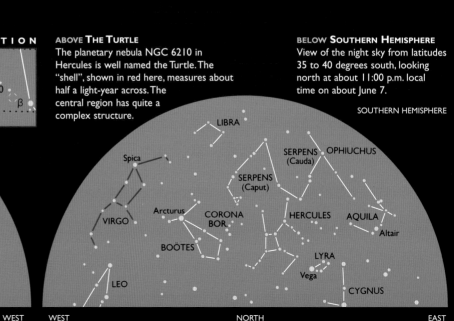

ABOVE **STING IN THE TAIL**
Scorpius was the creature that stung to death the mighty hunter Orion. The constellations were therefore placed on opposite sides of the heavens so that Orion disappears as Scorpius rises.

Scorpius, the Scorpion
a stunning southern constellation

Scorpius is one of the few constellations that really lives up to its name. Little imagination is needed to see the figure of a scorpion in its star pattern, with its deadly curved tail poised, ready to strike. Scorpius is the most spectacular of the constellations of the zodiac. One of the great delights of southern skies, it boasts eleven stars above the 3rd-magnitude, along with a host of naked-eye doubles and clusters.

LOCATION

RIGHT **NEAR THE SCORPION'S HEART**
The fine globular cluster M4 looks beautiful in small telescopes. The picture at near right is a ground telescope view of M4. At far right is a close-up view of the cluster by the Hubble Space Telescope showing a collection of white dwarf stars (circled).

RIVAL OF MARS

The brightest star in Scorpius is the 1st-magnitude Antares. Its name means "rival of Mars" because it is a noticeably reddish star that resembles the Red Planet, Mars.

Antares is a huge red supergiant, maybe as much as 700 times the size of the Sun. If it were where the Sun is in our Solar System, it would extend far beyond Mars into the asteroid belt. Antares is a very unstable star and pulsates, like most other supergiants. This makes it vary in brightness between about magnitude 9 and 1.2 about every five years.

SIGNIFICANT STARS

An arc of four stars—from Nu (ν) to Pi (π)—marks the scorpion's head. The constellation pattern no longer includes the creature's claws. They now form part of the neighboring constellation, Libra.

Nu is an interesting multiple star. It appears double in binoculars, and telescopes show that, additionally, each component is double, making a quadruple system.

Beta (β) is a double star, with bluish-white components. Antares itself has a close blue-white companion that orbits around it every 900 years. Zeta (ζ) is a naked-eye double. Zeta 1 is a striking reddish-orange star, while Zeta 2 is bluish-white. Mu (μ) and Omega (ω) are two other naked-eye doubles worth examining in binoculars.

CLASSIC CLUSTERS

Scorpius nestles in a rich part of the Milky Way and abounds in spectacular star clusters. Centered around Zeta 2 is the open cluster NGC 6231, whose young, hot white stars can easily be resolved in binoculars.

But the two best open clusters in Scorpius are M7 and M6, both naked-eye objects and located north of the scorpion's tail. Third-magnitude M7 is the brighter of the two as well as the largest, with almost double the diameter of the full Moon. M6 is known as the Butterfly Cluster because of the "open-wing" arrangement of its principal stars.

Scorpius also has some fine globular clusters. Close to Antares is M4, just on the limit of naked-eye visibility but easily picked up with binoculars. Somewhat fainter is M80, a very compact globular located roughly midway between Antares and Beta.

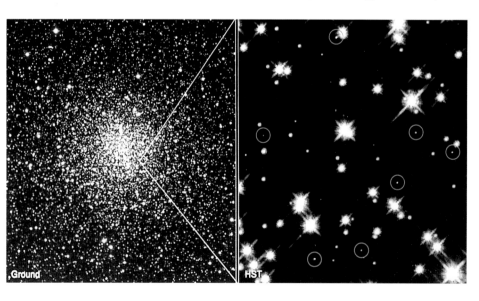

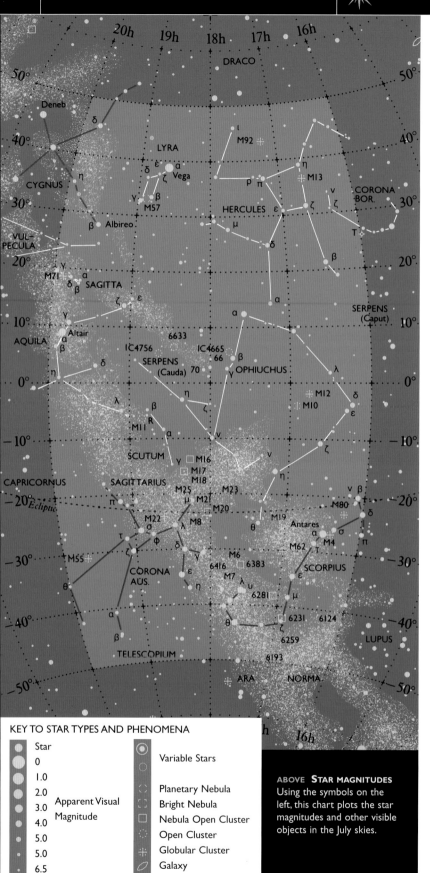

ABOVE STAR MAGNITUDES Using the symbols on the left, this chart plots the star magnitudes and other visible objects in the July skies.

KEY TO STAR TYPES AND PHENOMENA

Star
- 0
- 1.0
- 2.0 Apparent Visual
- 3.0 Magnitude
- 4.0
- 5.0
- 5.0
- 6.5

Variable Stars

Planetary Nebula
Bright Nebula
Nebula Open Cluster
Open Cluster
Globular Cluster
Galaxy

July Skies

Three brilliant stars appear high in the sky in the Northern Hemisphere, making up the celebrated Summer Triangle. Brightest is Vega in Lyra, which forms the triangle with Deneb in Cygnus and Altair in Aquila. In the Southern Hemisphere, of course, the three stars are features of winter and appear in low- to mid-sky. In both hemispheres, the Milky Way makes magnificent viewing on cloudless, moonless nights away from the light pollution of modern civilization.

Lyra, the Lyre

The musical instrument invented in Greek mythology by Hermes and played by Orpheus

One of the smallest constellations, Lyra is packed with a host of interesting objects, none more so than the breathtakingly beautiful M57, the Ring Nebula. This colorful "smoke ring" is a planetary nebula, puffed out by a dying Sun-like star. Sandwiched between Beta (β) and Gamma (γ), M57 can be seen as a faint disk in small telescopes, but only long exposures in large instruments will reveal its true technicolor splendor.

Brightest star Vega is a first-magnitude star that is the fifth brightest in the whole of the heavens. Blue-white in color and with a surface twice as hot as the Sun's, it lies about 25 light-years away.

The Double-Double

Close to Vega, outside the main star pattern, lies Epsilon (ε), well known as the "double-double" star. It is so called because it is a double star easily visible in binoculars. And when you look at its two components in a small telescope, each presents itself as a double too.

The four stars immediately south of Vega form a kind of parallelogram. Beta (β) is by far the most interesting. First, it is a double star, with yellowish and bluish components. Second, the brighter yellowish star of the pair is a kind of variable star we call an eclipsing binary.

Beta's two components circle around each other about once every 13 days, varying between the third and fourth magnitudes. They are very close together—probably less than 22 million miles (36 million km) apart. As a result, they are almost certainly egg-shaped, with gas streaming between them.

You can follow the variability of Beta by comparing its brightness with that of nearby Delta (δ) and Zeta (ζ), both of which are a steady magnitude 4.3. Incidentally, both these stars are doubles, which are easily separated through a small telescope.

Ophiuchus, the Serpent-Bearer

Identified with the Greek god of medicine Asclepius, who was one of Apollo's sons

This large constellation occupies mid-skies, straddling the celestial equator. It stretches between Hercules in the north and Scorpius in the south, where it enters a particularly rich region of the Milky Way. It splits the constellation Serpens in two—into Serpens Caput and Serpens Cauda, representing respectively the head and tail of the serpent.

To the eye, Ophiuchus is difficult to make out because it has no really bright stars. But it does contain some interesting clusters. In the south of the constellation, where it butts into the Milky Way, is the globular cluster M19. It lies roughly midway between Theta (θ) Ophiuchi and the brilliant orange Antares in the adjacent Scorpius.

Just north of Antares is Rho (ρ) Ophiuchi, which small telescopes reveal is a multiple star with three components. But the real glory of the region around Rho are the vast and colorful clouds of gas and dust that are a hotbed of star formation.

The interior of the constellation is noticeably empty of bright stars. But it does boast two fine globular clusters, M10 and M12. You can see both in binoculars, although you'll need a moderate-power telescope to resolve individual stars.

Sagittarius, the Archer

The key constellation featured in July See page 98

Scutum, the Shield

This is a relatively modern (1684) constellation, originally named Sobiesci's Shield after a Polish king

This tiny constellation just south of the celestial equator looks insignificant to the eye. But it lies near the edge of the Milky Way and is worth scanning via binoculars for its bright star clouds and nebulous regions.

Scutum's most notable feature is the open cluster M11, close to Beta (β), which is visible with binoculars. A small telescope reveals a V-shaped arrangement of stars, rather like a flight of ducks. This gives M11 its popular name, the Wild Duck cluster.

Serpens Cauda, the Serpent's Tail

The tail of the snake killed by Asclepius, which became a symbol of healing

This is the tail of Serpens that is separated from the head, Serpens Caput, by the constellation Ophiuchus. Perhaps the most interesting object in Cauda is the scattered open cluster M16. It is embedded in the particularly beautiful Eagle Nebula, which long exposures in telescopes show up well. In this nebula, the Hubble Space Telescope has taken one of its most dramatic pictures, named "the pillars of creation".

LOCATION

ABOVE LORD OF THE RINGS
The Ring Nebula in Lyra is the most magnificent of the ring-like planetary nebulae. The bright white-dwarf star at its center puffed off the ring of gases around 5,500 years ago.

BELOW NORTHERN HEMISPHERE
View of the night sky from latitudes 40–50 degrees north, looking south at about 11 p.m. local time on approximately July 7.

BELOW RIGHT SOUTHERN HEMISPHERE
View of the night sky from latitudes 35–40 degrees south, looking north at about 11 p.m. local time on about July 7.

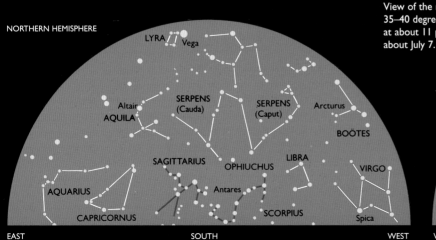

Sagittarius, the Archer, fiery constellation of the zodiac

Sagittarius is one of the richest constellations in the heavens, spanning the Milky Way. This milk-white band that represents a slice through our Galaxy is magnificent here. It is studded with brilliant star clusters, glittering nebulae, and thick star clouds. The densest regions mark the direction of the center of the galaxy, which lies some 25,000 light-years away. Sweeping this part of the night sky with binoculars takes your breath away.

FAR LEFT THE CENTAUR The Ancient Greeks pictured Sagittarius as a centaur, a strange creature that was half-human, half-horse. He is usually identified with Crotus, the inventor of the bow, who was the son of the pipe-playing god Pan. In the sky, he is aiming his arrow at Antares, the heart of Scorpius, the Scorpion.

LOCATION

BELOW TRIFID NURSERY In the Trifid Nebula, dark clouds of gas and dust are backlit dramatically by the intense radiation from a searing hot massive star beyond. In time, the radiation will strip away the dark clouds to reveal within them a host of newborn stars.

Alas for most observers in the Northern Hemisphere, Sagittarius never rises far enough above the horizon to be fully appreciated. How they envy astronomers "down under" in the Southern Hemisphere!

MESSIER BONANZA

When French astronomer Charles Messier drew up his catalog of star clusters and nebulae in 1781, he included more in Sagittarius than in any other constellation.

The nebulae in Sagittarius are absolutely stunning. The brightest, M8, is a 5th-magnitude object, and in clear skies can be glimpsed with the naked eye. Making a triangle with Mu (μ) and Lambda (λ), it is also known as the Lagoon Nebula. It is a fine object in a small telescope, but long-exposure photographs are needed to reveal its true beauty.

Just north of M8 and somewhat fainter (magnitude 7) is M20, known as the Trifid Nebula. You see it with binoculars as a small misty patch. With a telescope, you can make out the three dark dust lanes that give the nebula its name.

The third notable nebula in the constellation is M17, located some distance due north of Mu (μ). Easily visible with binoculars, it has roughly the same shape as the Greek capital letter omega (Ω), which is why it is also known as the Omega Nebula.

CLUSTER COLLECTION

Sagittarius also boasts an amazing variety of both open star clusters and globular clusters. There is one open cluster, M18, just south of the Omega Nebula, and south of this again is M24. Neither a nebula nor a cluster, M24 is a collection of bright stars in the Milky Way. It is a naked-eye object, and is also known as the Small Sagittarius Star Cloud.

Making a triangle with M24 and Mu (μ) is the fine open cluster M23. You can see it with binoculars, and resolve some of its 100 or so stars with a small telescope.

GREAT GLOBULARS

Sagittarius boasts more than 20 globular clusters. Of them all, M22 is the finest, being a globe-shaped mass of literally millions of stars. Located north of Lambda (λ), it is just visible to the naked eye under ideal conditions and looks great with binoculars and a small telescope. It is the third brightest globular, after 47 Tucanae and Omega Centauri.

Three other globulars can be spotted along an imaginary line joining Epsilon (ε) and Zeta (ζ). They are M69, M70, and M54, which are all about the 8th-magnitude.

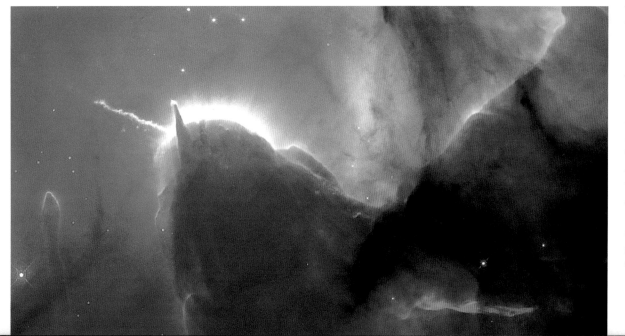

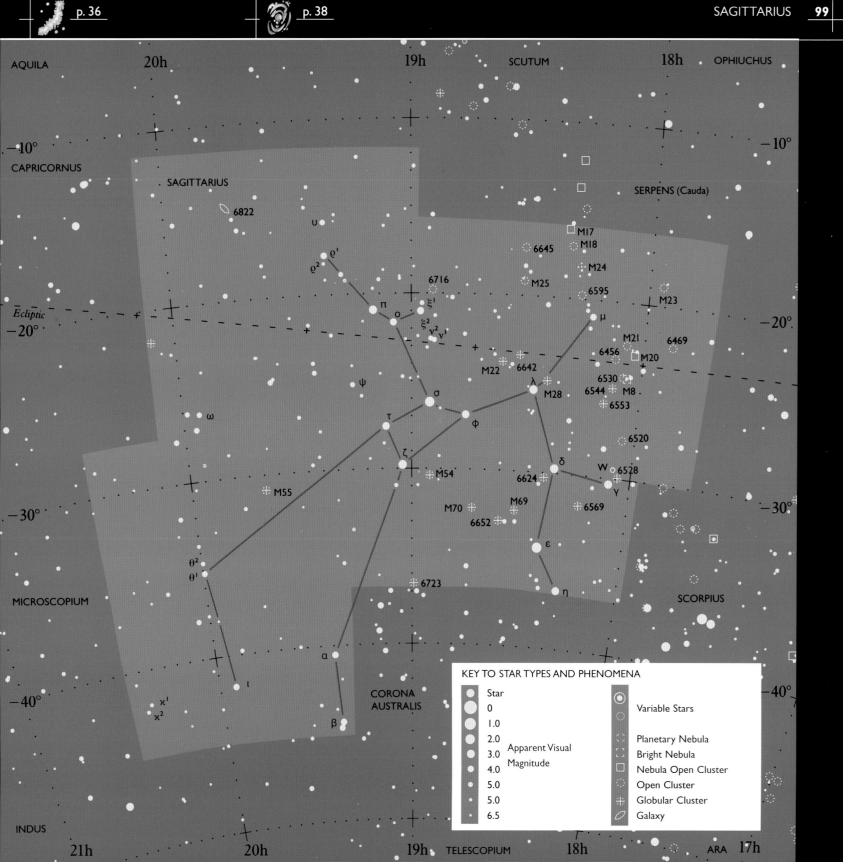

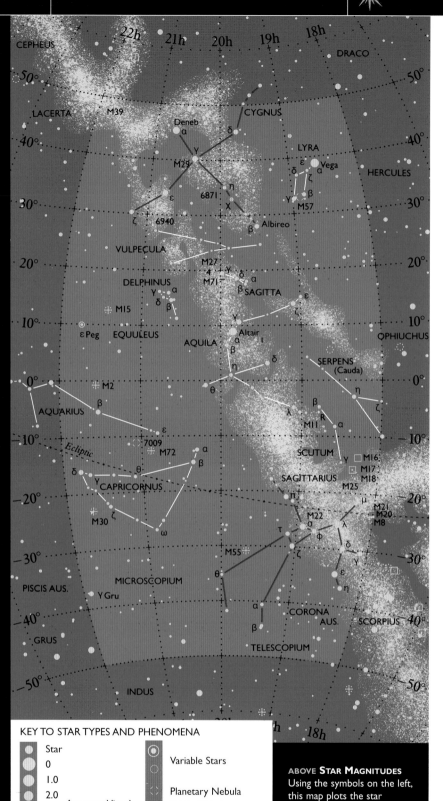

August Skies

This month, the Summer Triangle of brilliant white—Deneb, Vega, and Altair—straddles the meridian, set against the stunning backdrop of the Milky Way. For northern observers this is a good month to see Sagittarius, low in the south, close to the meridian. For southern observers this is the best month to sample the delights of Cygnus.

AQUILA, THE EAGLE

The bird that carried thunderbolts for Zeus, the king of the gods

Set in the Milky Way, spanning the celestial equator, and appearing on the meridian this month, Aquila is a fine constellation. Its lead star Alpha (α), or Altair, forms one corner of the Summer Triangle. At magnitude 0.8 and brilliant white, Altair is one of the closest bright stars, located only 16 light-years away.

Flanking Altair are the noticeably orange Gamma (γ) and Beta (β), the set of three being easy to recognize. South of Beta is another line of three stars Theta (θ), Eta (η), and Delta. Eta is the most interesting because it is one of the brightest Cepheid variables. It varies between the 3rd- and 4th-magnitudes in just over seven days. Its progress can be followed by comparison with 3rd-magnitude Theta and 4th-magnitude Delta.

The curve of stars around Lambda (λ) at the southern end of the constellation act as useful pointers to M11, the Wild Duck open star cluster in neighboring Scutum.

CAPRICORNUS, THE SEA GOAT

Myths link this fish-tailed goat with the Greek god Pan

One of the fainter constellations of the zodiac, Capricornus is not easy to recognize. Its brightest stars are all about 3rd-magnitude. Alpha (α) is a wide double star, separated by the naked eye and easily seen in binoculars. Small telescopes will find that each component is itself double. Beta (β) is also a double, visible in binoculars.

The brightest deep-sky object is the globular cluster M30, just east of Zeta (ζ). Small telescopes reveal it, but larger instruments are needed to make out individual stars.

CYGNUS, THE SWAN

The key constellation featured in August See page 102

DELPHINUS, THE DOLPHIN

The fish sent by the sea god Poseidon to fetch the beautiful nymph Amphitrite to be his bride

Delphinus is a small but distinctive constellation, which you can easily imagine to be a leaping dolphin. Appearing in the same field in binoculars, its bright stars form a compact group that looks almost like a star cluster. The quadrilateral formed by the head stars is known as Job's Coffin. Among them, Gamma (γ), a fine double, separates into gold components when viewed through small telescopes.

ABOVE STAR MAGNITUDES Using the symbols on the left, this map plots the star magnitudes and other visible objects in August skies.

SAGITTA, THE ARROW

An arrow shot by the hero Hercules, otherwise known as Cupid's arrow

The third smallest constellation, Sagitta has a distinctive arrow shape. It sits in the Milky Way and is readily located north of Altair in neighboring Aquila. The four main stars make really nice viewing through binoculars against the dense star fields of the Milky Way. Between Delta (δ) and Gamma (γ) is a globular cluster, M71, seen best via small telescopes.

VULPECULA, THE FOX

A modern constellation, introduced in the 17th century

No star in this faint constellation is brighter than 4th-magnitude. But, being in the Milky Way, the region is worth scanning with binoculars. The most interesting deep-sky object is M27, or the Dumbbell Nebula. One of the brightest planetary nebulae, it is located due north of Gamma (γ) Sagittae in the adjacent constellation.

LOCATION

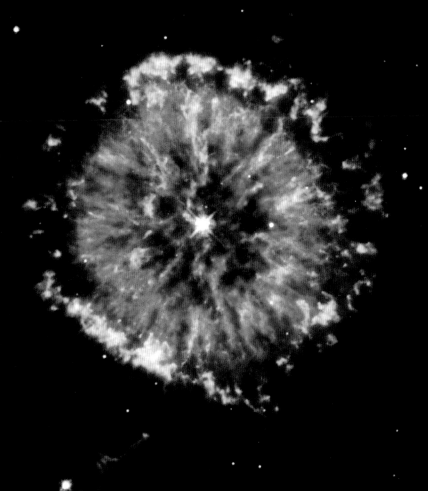

LEFT THE EAGLE'S EYE
A stunning Hubble Space Telescope image of NGC 4751 in Aquila. It is a planetary nebula, which looks rather like an eye with a multicolored iris. The brilliant white dwarf star at the center could be as hot as 250,000°F (140,000°C).

BELOW NORTHERN HEMISPHERE
View of the night sky from latitudes 40 to 50 degrees north, looking south at about 11:00 p.m. local time on about August 7.

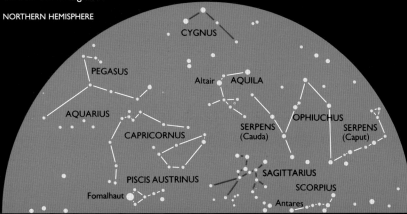

BELOW SOUTHERN HEMISPHERE
View of the night sky from latitudes 35 to 40 degrees south, looking north at about 11:00 p.m. local time on about August 7.

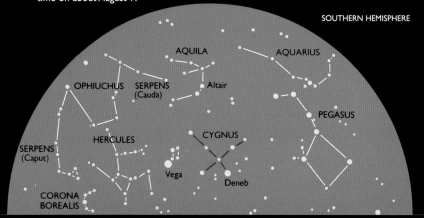

Cygnus, the Swan
the distinctive Northern Cross

One of the most beautiful northern constellations, Cygnus looks impressively like a bird in flight, with widespread wings and long neck outstretched. It seems to be flying south along the Milky Way, pursuing the eagle, represented by the constellation Aquila. Both "birds" boast 1st-magnitude stars, which form two corners of the celebrated Summer Triangle.

ABOVE ZEUS IN DISGUISE
When Zeus went philandering he usually disguised himself to get close to the object of his desires. He assumed the disguise of a swan when he pursued Leda, Queen of Sparta. The result of the liaison was an egg, out of which hatched the twins Castor and Pollux; and Helen, who would become the beautiful Helen of Troy, with the "face that launched a thousand ships."

It is Cygnus's lead star Deneb that forms one corner of the Summer Triangle. At magnitude 1.3 it is an exceptionally luminous supergiant, with the light output of 60,000 Suns. It lies much farther away than the other 1st-magnitude stars, at a distance of more than 1,800 light-years.

DAZZLING DOUBLES

Cygnus's head star, Beta (β), or Albireo, lies at the opposite end of the constellation. It is actually the faintest of the four stars that make up the main cross pattern. But Albireo is a gem, being a really lovely double, separated via small telescopes into blue and gold components.

Omicron (o) is an easy binocular double, with reddish and bluish components. But the most famous double is 61, beneath the swan's right wing, forming a triangle with the brighter Sigma (σ) and Tau (τ). 61 Cygni was the first star to have its distance measured accurately—by the German astronomer Friedrich Bessel in 1838.

VARIOUS VARIABLES

Several Mira-type long-period variables are worth studying in the constellation. Some, such as U and R Cygni, come within binocular range at maximum, but fade beyond their range at minimum.

The most impressive variable is Chi (χ), in the swan's neck. At maximum, Chi reaches 3rd-magnitude and rivals nearby Eta (η) in brightness. But it falls below 14th-magnitude at minimum, beyond small telescope range, over a period of 407 days. It thus displays one of the greatest brightness changes among variables we know.

W Cygni, near Rho (ρ) east of Deneb, is another interesting variable. Its change of brightness, between 5th- and 8th-magnitude, can be followed throughout with binoculars. It varies over a period of about four months.

CLUSTERS AND NEBULAE

Being within the Milky Way, Cygnus abounds in deep-sky objects. M39 and M29 are two fine open clusters. Fifth-magnitude M39 is the brightest; it is found in the tail end of the constellation, north of Rho. Located south of Gamma (γ), M29 is two magnitudes fainter.

The Milky Way east of Deneb brightens noticeably, and binoculars and small telescopes reveal a glowing nebula, NGC 7000.

Long-exposure photographs show a distinctive shape that gives it the name, the North America Nebula.

South of Epsilon (ε), around the star 52, small telescopes show glowing filaments of glowing gas, forming the Veil Nebula. They are part of a larger region known as the Cygnus Loop, which is a supernova remnant.

LOCATION

ABOVE THE VEIL NEBULA
This delicate tapestry of glowing gas is known as the Veil Nebula (NGC 6992). It is part of a much larger luminous ring named the Cygnus Loop, which is the remains of a supernova explosion that took place some 30,000 years ago.

LOCATION

RIGHT RIFT IN CYGNUS
A dense star field in the Milky Way in Cygnus. The brightest star near the center is the 1st-magnitude Deneb. Close by is a nebulous region including the North America Nebula. At upper left in the picture is a "hole" in the Milky Way, actually a massive dark nebula known as the Cygnus Rift.

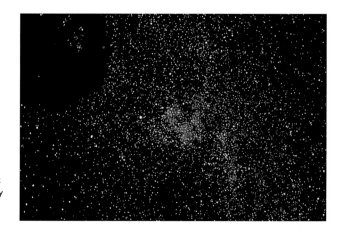

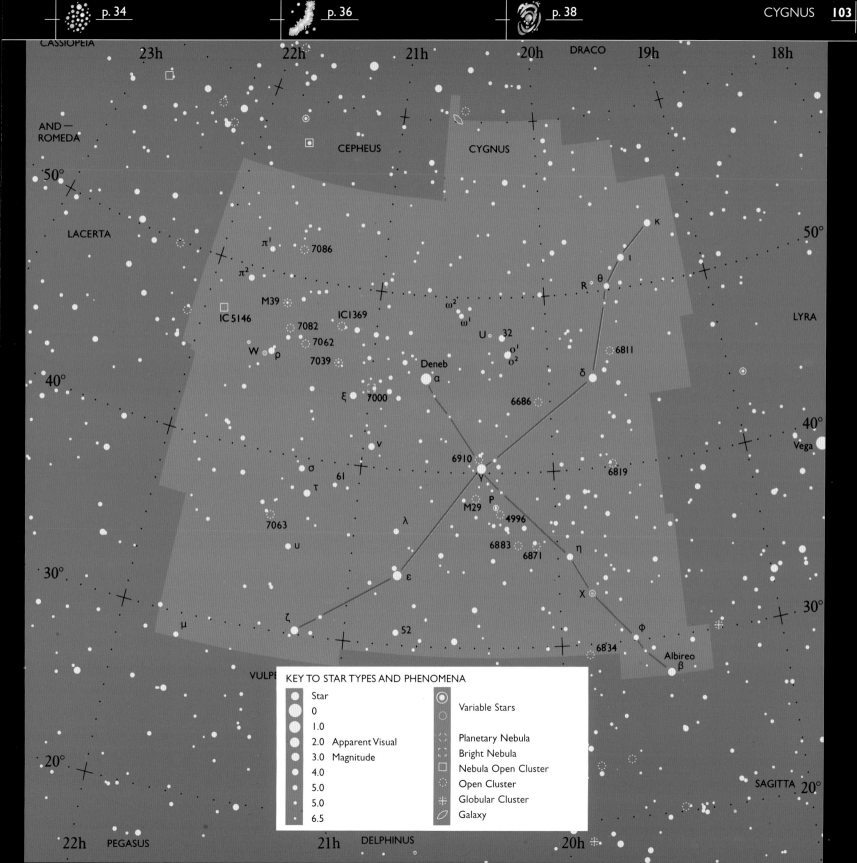

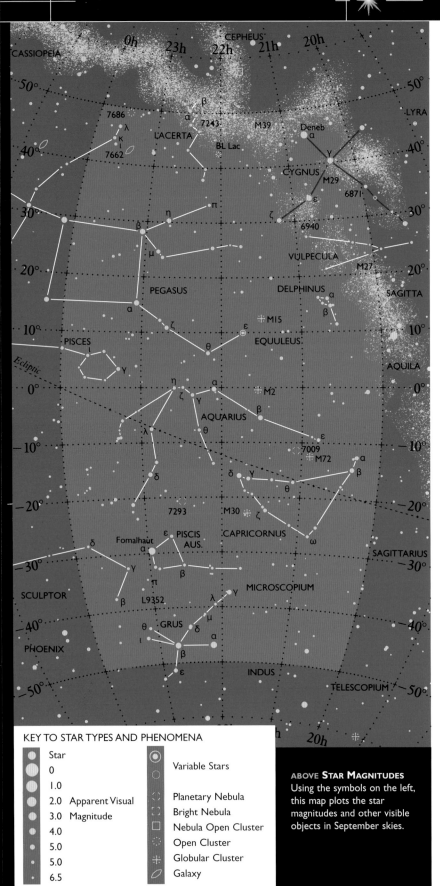

ABOVE STAR MAGNITUDES Using the symbols on the left, this map plots the star magnitudes and other visible objects in September skies.

September Skies

The bright Summer Triangle that has dominated the skies for the past few months is moving west, while the unmistakable Square of Pegasus is climbing fast from the east. This signals the rapid approach of Fall in the Northern Hemisphere and of Spring in the Southern. The Milky Way has also moved from center stage, making the skies duller by comparison. Fainter constellations have taken its place.

AQUARIUS, THE WATER-BEARER

The handsome youth Ganymede, cup-bearer to the gods, pouring water from a jar

This sprawling constellation of the zodiac has leading stars of only about 3rd-magnitude and is not easy to define. Its two brightest stars Alpha (α) and Beta (β) can probably best be found by extending a line from Alpha Andromedae and Alpha Pegasi in the Square of Pegasus.

The stars immediately east of Alpha form part of the mouth of the water jar, and the water trickles south along the line of stars tumbling toward the mouth of the Southern Fish, Piscis Austrinus.

Due north of Beta, and making a right-angled triangle with Alpha, is the relatively bright globular cluster M2. It is visible as a misty patch in binoculars and resolved into stars in small telescopes. Just south of Epsilon (ε) is another fairly bright globular cluster, M72, best seen in small telescopes.

Aquarius also boasts two fine planetary nebulae. One, NGC 7009, lies close to M72. Seen through larger telescopes, it looks rather like a ringed planet, which is why it is called the Saturn Nebula. The other planetary nebula, NGC 7293, is located in the south of the constellation. It is one of the nearest planetary nebulae to us, at a distance of about 300 light-years. It is large but quite faint, and larger telescopes are needed to bring out its spiral structure, which gives it the name, the Helix Nebula.

GRUS, THE CRANE

One of the southern birds; a modern constellation, introduced in the 17th century

This constellation is located south of Piscis Austrinus, with a long, curving line of stars marking the neck and body line. At magnitude 1.7, Alpha (α) is slightly brighter than Beta (β), Alpha being bluish-white and Beta distinctly orange. Lambda (λ) and Iota (ι) also have an orange hue. Both Delta (δ) and Mu (μ) are naked-eye doubles.

LACERTA, THE LIZARD

A modern constellation, introduced in the 17th century

This faint northern constellation is sandwiched between Cassiopeia and Cygnus, and extends into the Milky Way. For this reason it is worth sweeping with binoculars. NGC 7243 is an open cluster to the west of Alpha (α).

Of great astronomical consequence, but too faint to be seen in most amateur instruments, is the object BL Lacertae. It is a galaxy that changes noticeably in brightness in a relatively short period of time. Astronomers think that it is one of a class of active galaxies known as blazars, which are similar to quasars.

SEPTEMBER SKIES

PISCIS AUSTRINUS, THE SOUTHERN FISH

The parent of the two fish represented in the heavens by the constellation Pisces

This small constellation is dominated by a single bright star, 1st-magnitude Fomalhaut (Fish's Mouth). No other stars are above 4th-magnitude. Although Fomalhaut ranks as only the eighteenth brightest star, it is easy to spot because it occurs in a relatively empty region of the heavens.

Among the other stars, Beta (β) is a double, resolvable in small telescopes. But perhaps the most interesting object in the constellation is a star called Lacaille 9352. Of 8th-magnitude, it lies less than 12 light-years from Earth and is notable because of its pronounced proper motion—its motion across our line of sight. This motion is easily discernible in photographs taken one year apart.

LOCATION

LEFT COMETARY KNOTS
Comet-like blobs of gas are escaping from the inner edge of the ring that forms the Helix Nebula in Aquarius. Each one measures about twice the width of our Solar System. And their tails stretch for 100 billion miles (160 billion km).

BELOW LEFT NORTHERN HEMISPHERE
View of the night sky from latitudes 40 to 50 degrees north, looking south at about 11:00 p.m. local time on about September 7.

BELOW SOUTHERN HEMISPHERE
View of the night sky from latitudes 35 to 40 degrees south, looking north at about 11:00 p.m. local time on about September 7.

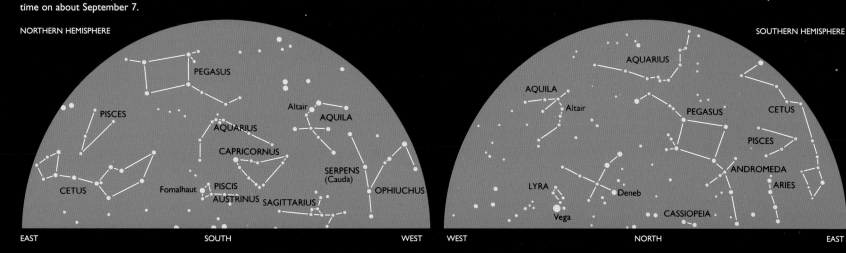

October Skies

October skies are dominated by the unmistakable Square of Pegasus, which sits squarely on the meridian. With few bright stars, the surrounding "watery" constellations like Pisces, Aquarius, and Cetus are harder to trace. October is a good month in both hemispheres for observing one of our closest galactic neighbors, M31 in Andromeda.

Andromeda

The key constellation featured in October
See page 108

Pegasus, the Flying Horse

Born from the body of Medusa, whom Perseus slew, Pegasus was tamed by Bellerophon, who rode it to kill the monster, Chimaera

The seventh largest constellation, Pegasus lights up Fall skies in the Northern Hemisphere and Spring skies in the Southern. Its four main stars make up one of the most distinctive shapes in the heavens—the great Square of Pegasus.

Pegasus shares one of the stars in the Square with neighboring Andromeda—Alpha (α) Andromedae, also named Alpheratz. At magnitude 2.1, it is the brightest of the Square stars.

Beta (β) in the Square varies slightly in brightness between magnitudes 2.3 and 2.7. It is a red giant and has a definite reddish hue, compared with the pure white Alpha (α). Of the stars outside the Square, the brightest is Epsilon (ε), also known as Enif. It also has a reddish tinge. It is a double star, visible in binoculars.

Among deep-sky objects in Pegasus, M15 is outstanding. It is a compact globular cluster easily seen in binoculars. It lies in line with Theta (θ) and Epsilon (see map on page 105 for location).

The apparently insignificant star 51, north of Alpha, is historically important. It was the first ordinary star found (in 1995) to have a planet circling around it. This extrasolar planet appears to be about half the size of Jupiter.

Phoenix, the Phoenix

A 17th-century creation; the bird that was reborn from its own ashes

Another of the "southern birds," Phoenix is a rather barren constellation, with only Alpha (α) being 2nd-magnitude. Beta (β) is a fine double star, with equally bright yellowish components. Zeta (ζ) is another double, and its brightest component is an eclipsing binary, like Algol in Perseus. It dips briefly in brightness every 40 hours.

Pisces, the Two Fish

The fish that Venus and her son Cupid turned into to escape from the clutches of the monster Typhon

This large and faint constellation of the zodiac is formed by a string of stars little brighter than 4th-magnitude. One of the fish is located just south of the Square of Pegasus and is marked by an oval of stars named the Circlet. The other fish appears a long way off, south of Beta (β) Andromedae. Lines of faint stars emanating from Alpha (α) join the two fish, but they are not easy to make out.

ABOVE STAR MAGNITUDES Using the symbols on the left, this map plots the star magnitudes and other visible objects in October skies.

KEY TO STAR TYPES AND PHENOMENA

Star
- 0
- 1.0
- 2.0 Apparent Visual
- 3.0 Magnitude
- 4.0
- 5.0
- 5.0
- 6.5

- Variable Stars
- Planetary Nebula
- Bright Nebula
- Nebula Open Cluster
- Open Cluster
- Globular Cluster
- Galaxy

OCTOBER SKIES

Larger telescopes reveal that Alpha is a binary star, while smaller ones resolve the two yellow components of the double Zeta (ζ) farther west.

The brightest of several galaxies in Pisces is the face-on spiral M74, which is located close to Eta (η). Small telescopes spy it as a small, round misty patch.

SCULPTOR, THE SCULPTOR

A modern constellation, introduced in the 18th century

This faint constellation can be found easily by looking due east of Fomalhaut in Piscis Austrinus. It is notable, like its eastern neighbor Fornax, mainly for its galaxies. One of the finest is NGC 253, located north of Alpha (α) and visible with binoculars. Telescopes reveal that this spiral galaxy presents itself nearly edge-on.

LOCATION

ABOVE CENTER STARS
A view of the globular cluster M15 in Pegasus taken by the Hubble Space Telescope. The inset is a magnified image of the center of the cluster, showing individual stars. The bluish-white ones are the hottest; the reddish-orange ones, the coolest.

BELOW NORTHERN HEMISPHERE
View of the night sky from latitudes 40 to 50 degrees north, looking south at about 11:00 p.m. local time on about October 7.

BELOW RIGHT SOUTHERN HEMISPHERE
View of the night sky from latitudes 35 to 40 degrees south, looking north at about 11:00 p.m. local time on about October 7.

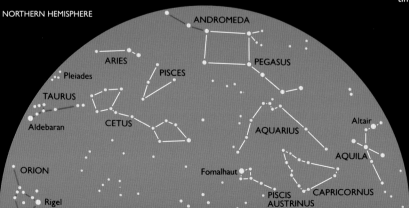

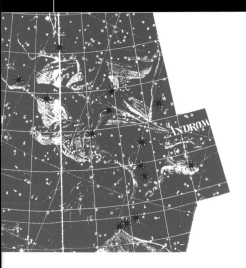

Andromeda
home of the Great Spiral

This large northern constellation is easy to find because it is linked to one of the most distinctive patterns in the heavens—the Square of Pegasus. Andromeda is famous, not for its stars, but for a little misty patch that Charles Messier listed as M31. It used to be called the Great Nebula in Andromeda, but we now know it is a neighboring galaxy.

ABOVE THE SACRIFICED DAUGHTER
Andromeda was the beautiful daughter of King Cepheus and the vain Queen Cassiopeia. She upset the sea god Poseidon by boasting about her beauty, and he ordered the King to sacrifice Andromeda to the sea monster Cetus. Chained to the rocks and about to be devoured, Andromeda was rescued by the hero Perseus, who happened by. Later, they were married.

LOCATION

BELOW BRIGHT SATELLITE
A close-up view of M32, in false color, taken by the Hubble Space Telescope. It is the closest of the two satellites of the Andromeda galaxy, M31. Notice how the brightness increases toward the center, where there is thought to be a supermassive black hole with the mass of three million Suns.

Andromeda's brightest star, Alpha (α), or Alpheratz, joins the constellation to the Square of Pegasus. Along with Beta (β) and Gamma (γ), it has a magnitude of 2.1. Gamma is a beautiful double, separated via small telescopes into a golden yellow primary and bluish-green companion.

Among the variable stars in the constellation is R Andromedae, located close to 5th-magnitude Rho (ρ). It is a Mira-type variable, which becomes as bright as 6th-magnitude but falls to below 15th-magnitude over a period of 409 days. It is a pulsating red giant star, which is visibly very red.

THE GREAT SPIRAL

The Andromeda Galaxy, M31, is one of the foremost objects in the whole night sky. It is a spiral galaxy, like our own in structure, but much bigger. With a diameter of around 150,000 light-years, it measures half as big again as our own Galaxy and could contain as many as 400 billion stars. It is the largest galaxy by far in our local galaxy cluster (The Local Group) of about 30 galaxies.

Of about 4th-magnitude, the Andromeda Galaxy is easily seen with the naked eye as a misty patch close to the star Nu (ν). It lies about 2.5 million light-years away, which makes it easily the most distant object we can see in the heavens with the naked eye.

In binoculars, the galaxy shows up as a distinct oval, while larger instruments show pronounced dark dust lanes and begin to resolve individual stars. M31 has two companion, or satellite, galaxies—the elliptical M32 and the spiral M110 (NGC 205). M32 is visible with binoculars, but M110 requires a small telescope.

OTHER DEEP-SKY OBJECTS

Two other galaxies are worth looking for in Andromeda. One is NGC 891, found east of Gamma. It is a classic edge-on spiral, seen in larger telescopes to be bisected by a dark dust lane. Next to Beta is an elliptical galaxy, NGC 404.

Following an arc through Lambda (λ), Kappa (κ), and Iota (ι) leads you to quite a bright planetary nebula NGC 7662. In small telescopes it looks like a fuzzy greenish star, but larger instruments will reveal its donut-shaped disk and central white-dwarf star.

In the east of the constellation, south of Upsilon (υ), is the star 56, and close to it is a widely scattered open cluster, NGC 752, visible via binoculars.

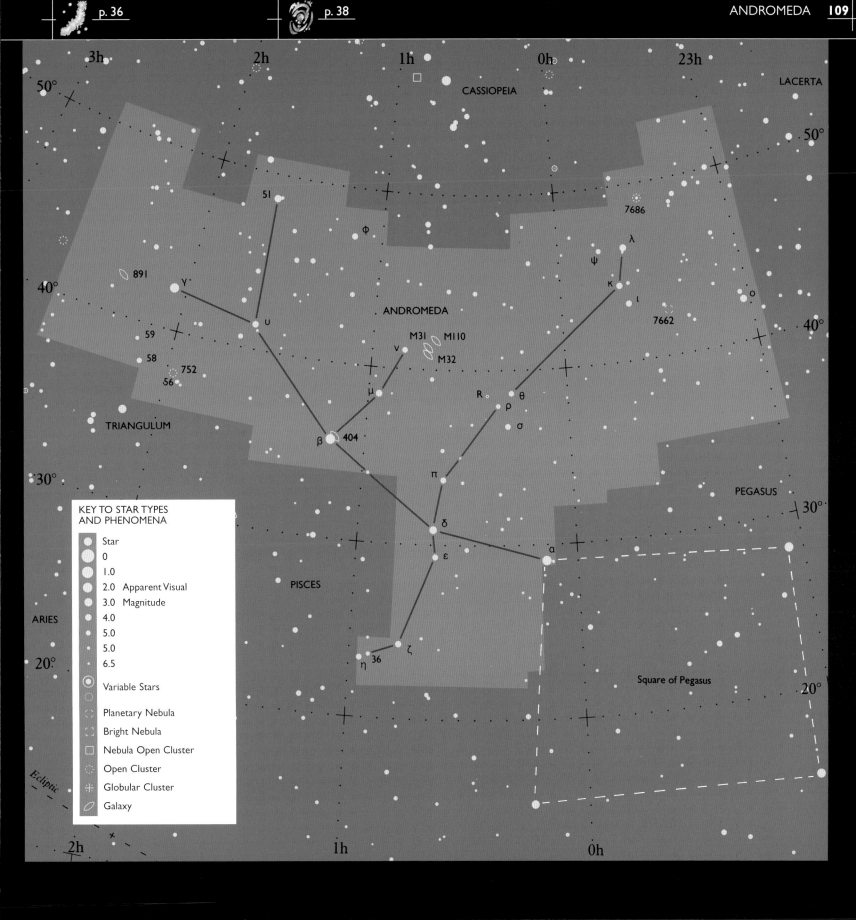

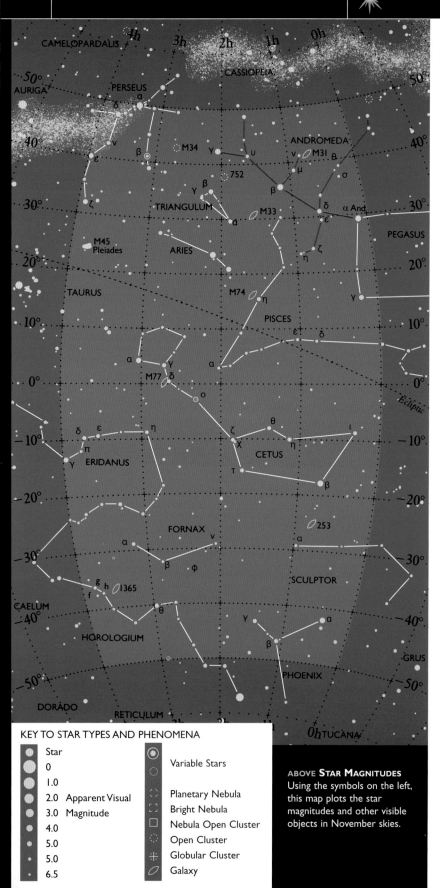

ABOVE STAR MAGNITUDES Using the symbols on the left, this map plots the star magnitudes and other visible objects in November skies.

November Skies

With Pegasus descending in the west, much of the sky is still dull because of the faint "watery" constellations like Pisces, Cetus, and Eridanus. But in the east the skies are beginning to brighten as Taurus and Orion begin climbing higher. Taurus's two bright clusters, the Hyades and Pleiades, are becoming well-placed for observation in both hemispheres.

ARIES, THE RAM

The ram with the Golden Fleece, which was sought by Jason and the Argonauts

This small constellation of the zodiac has just two bright stars, 2nd-magnitude Alpha (α) and the slightly fainter Beta (β). The two are found roughly midway between the Pleiades cluster and the Square of Pegasus. Gamma (γ), to the south of Beta, is an easy double for small telescopes, being resolved into two equally bright blue-white stars.

CETUS, THE WHALE

Generally pictured as a sea monster rather than a whale, it was the creature to which Andromeda was to be sacrificed, before Perseus fought and killed it

The fourth largest constellation, Cetus spans the celestial equator. But it is not easy to trace because of its lack of bright stars. It is probably best found by following the line of stars from Aldebaran in the southern part of Taurus.

Only Alpha (α), in the whale's head, and Beta (β), in the tail, are above 3rd-magnitude. At magnitude 2, Beta is half a magnitude brighter than Alpha.

Alpha is a wide binocular double. It is a red giant, and does look noticeably red. The adjacent Gamma (γ) is a binary, whose bluish-white and yellow components can be separated in a telescope. Third-magnitude Tau (τ), in the tail region, is interesting because it is one of the nearest Sun-like stars, 11.9 light-years away.

Historically, the most interesting star in the constellation is Omicron (ο), also called Mira ("the wonderful"). It is the prototype long-period variable star, a huge red giant that pulsates and varies widely in brightness over a period of months or a year or more. Dutch astronomer David Fabricus first noted its variability in 1596.

Mira Ceti, found roughly halfway between Alpha and the distinctive pair Zeta (ζ) and Chi (χ), varies between the 3rd- and 10th-magnitude every 11 months or so. Its variation can be followed in binoculars.

Among deep-sky objects, the galaxy M77 can be found near 4th-magnitude Delta (δ)—seen in binoculars as a hazy patch. Telescopes show it is a face-on spiral, with wide-open arms. M77 is an active galaxy with a particularly bright center, a type known as a Seyfert.

TRIANGULUM, THE TRIANGLE

The Greeks orginally named this after the Greek letter, capital delta; the Romans identified it with the island of Sicily

One of the tiniest constellations, Triangulum is well named, with its three brightest stars forming a neat triangle. In the triangle, Beta (β) is the brightest star at magnitude 3; Gamma (γ) is magnitude 4, while Alpha (α)

NOVEMBER SKIES

has a brightness halfway between them. Gamma and the two stars that flank it form a nice group best observed with binoculars.

Triangulum's most outstanding feature is the beautiful face-on spiral galaxy M33, located east of Alpha in the direction of the star Beta in neighboring Andromeda.

On a really dark night, observers with good eyesight can just make it out, but it is easy to find in binoculars, appearing as a small misty patch. Larger amateur telescopes are able to resolve the wide-open arms of the galaxy, which forms part of the Local Group that includes our own Galaxy and the Andromeda Galaxy.

BELOW Northern Hemisphere
View of the night sky from latitudes 40 to 50 degrees north, looking south at about 11:00 p.m. local time on about November 7.

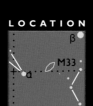

LOCATION

LEFT The Third Spiral
Ground telescope view of M33 in Triangulum, one of only three spiral galaxies in the Local Group. It is less than half the size of our Galaxy, with a diameter of about 40,000 light-years. It lies about 2.5 million light-years away.

LOCATION

ABOVE M33's Stellar Nursery
In the outer reaches of M33 lies the luminous gas cloud NGC 604, pictured spectacularly here by the Hubble Space Telescope. It is one of the largest star-forming regions we know of. The gas is being lit up by the intense radiation given out by clutches of hot, newborn stars.

BELOW Southern Hemisphere
View of the night sky from latitudes 35 to 40 degrees south, looking north at about 11:00 p.m. local time on about November 7.

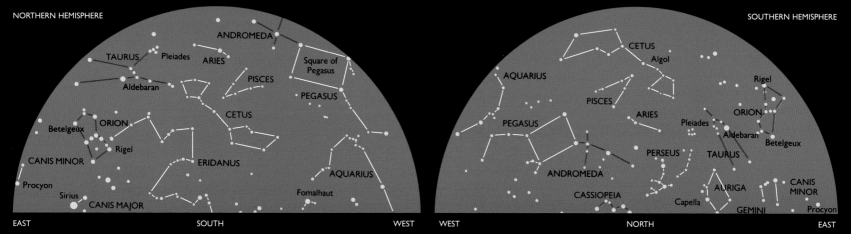

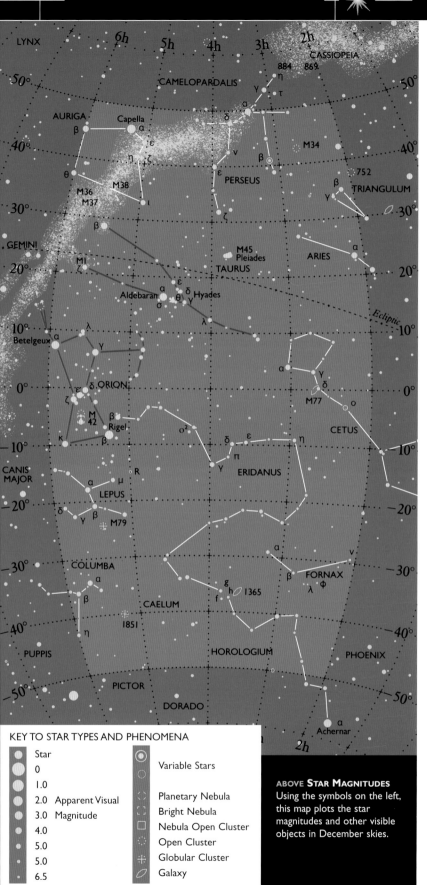

ABOVE STAR MAGNITUDES Using the symbols on the left, this map plots the star magnitudes and other visible objects in December skies.

December Skies

In December, the skies are shaping up to become among the most spectacular of the year for stargazing. They are studded with brilliant constellations and a surfeit of beacon stars. Taurus has climbed to the meridian, with Orion not far behind. It is eastern skies that boast the brightest stars, including the Dog Star Sirius, Rigel, Castor and Pollux, Capella, and Procyon.

ERIDANUS

A meandering river, which the Babylonians associated with the Euphrates, and the Egyptians with the Nile

This constellation is exceptionally long and winds its way south like a river, from near the celestial equator and deep into southern polar regions. Its second brightest star Beta (β), about 3rd-magnitude, near Orion's Rigel, marks the source of the river. Alpha (α), or Achernar, of magnitude 0.5, marks the river's mouth.

Heading "upstream" from Achernar, Theta (θ) is the first interesting star. It is one of the finest doubles in southern skies. Small telescopes separate it into twin white stars. Near the source of the river, Omicron 2 (o) is also double. Small telescopes show it has a 10th-magnitude companion, which happens to be a white dwarf. In fact it is one of the easiest white dwarfs to see from Earth. Larger instruments reveal that it forms a binary with another tiny star, this time a red dwarf.

Delta (δ) and Pi (π) both look distinctly orange through binoculars. They form a triangle with Epsilon (ε). This star is notable because, at a distance of 10.8 light-years, it is the third nearest naked-eye star to us (after Alpha Centauri and Sirius).

FORNAX, THE FURNACE

A modern constellation, introduced in the 18th century

This small, faint constellation has only one star, Alpha (α), above 4th-magnitude. Small telescopes show that this star is a binary. Small telescopes also begin to reveal the riches this constellation possesses.

Zeroing in on the twin stars Lambda (λ), reveals a dense star cluster nearby. Larger instruments show that it is an irregular dwarf galaxy, called the Fornax System.

On the eastern edge of the constellation, near the stars g, f, and h in Eridanus, small telescopes also find the galaxy NGC 1365. At about 10th-magnitude, it is one of the brightest members of the Fornax Cluster of galaxies. Larger telescopes show it to be a barred-spiral galaxy. A somewhat fainter galaxy in the same region is NGC 1316, an elliptical galaxy that is a powerful radio source, known as Fornax A.

PERSEUS

One of the great Greek heroes, whose feats included killing the Gorgon Medusa and the monster Cetus

This bright northern constellation straddles the Milky Way. Leading star Alpha (α), or Mirfak, looks bright at magnitude 1.8. It is surrounded by a myriad of faint stars, lovely via binoculars.

One highlight of Perseus is the famous Double Cluster, a pair of glorious open clusters, visible to the keen eye on dark nights. Also called the Sword Handle, they are quite easily located just to the north of Eta (η). On star charts they are usually identified as h and Chi Persei, or NGC 869 and 884. Another naked eye cluster is M34, to the west of Beta (β).

Beta is the other showpiece object in Perseus, also known as Algol and the Winking Demon. For most of the time, Beta shines steadily at about magnitude 2. But every 2 days 21 hours, it drops to around magnitude 3.3 for about 10 hours, before regaining its original brightness. Algol is a classic example of the two-star system known as an eclipsing binary.

TAURUS THE BULL

The key constellation featured in December
See page 114

LOCATION

RIGHT SOUTHERN STARBURSTS
In the southern constellation Horologium (the Clock), the Hubble Space Telescope has peered into the heart of the barred-spiral galaxy NGC 1512. It has spied a ring of new star formation, which astronomers call a starburst.

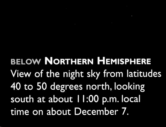
BELOW NORTHERN HEMISPHERE
View of the night sky from latitudes 40 to 50 degrees north, looking south at about 11:00 p.m. local time on about December 7.

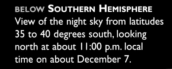
BELOW SOUTHERN HEMISPHERE
View of the night sky from latitudes 35 to 40 degrees south, looking north at about 11:00 p.m. local time on about December 7.

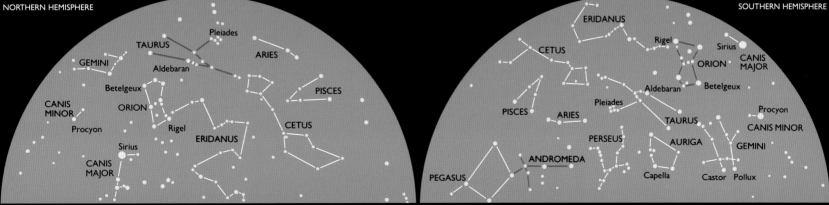

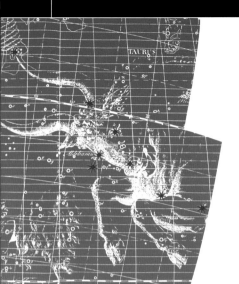

Taurus, the Bull
with red eye and thrusting horns

This splendid constellation of the zodiac boasts two of the best open clusters in the heavens—the Hyades, gathered around 1st-magnitude Aldebaran, and the Pleiades, farther out. Both are glorious naked-eye objects. Aldebaran marks the eye of the bull. It is a red giant and, like the nearby Betelgeuse in Orion, has a distinctive reddish-orange hue.

ABOVE BULL TO WOO
In amorous pursuit of the beautiful Europa, daughter of the king of Phoenicia, Zeus turned himself into a magnificent snow-white bull. When Europa climbed on his back, he swam to the island of Crete and seduced her. One of the children she bore was King Minos, who built the famed palace of Knossus, noted for the bull games played there.

LOCATION

RIGHT GALAXIES IN A SPIN
These two galaxies in Taurus collided millions of years ago. Now they are orbiting each other at a speed of 620,000 miles an hour (1,000,000 km/h) or more. Matter is being "piped" between the galaxies, visible as a dark line.

Aldebaran, Alpha (α) Tauri, is easy to find because it is in line with the three stars that make up Orion's Belt—and the Pleiades lies farther away along the same line. Its magnitude is 0.8 that compares with 1.6 for Beta (β), also called Al Nath. Beta marks the tip of one of the bull's horns. Third-magnitude Zeta (ζ) marks the tip of the other. Both Beta and Zeta lie on the edge of the Milky Way.

THE HYADES

The Hyades open cluster is arranged in a rough V-shape from around Aldebaran. But Aldebaran is not part of it. Whereas the Hyades lies about 130 light-years away, Aldebaran is only half this distance. Theta is the brightest star in the Hyades, at about magnitude 3.5. It is a double, with a white primary and a reddish-orange companion. Altogether, the Hyades may contain as many as 200 stars.

THE SEVEN SISTERS

North-west of the Hyades is the even more spectacular open cluster called the Pleiades. It is also known as the Seven Sisters, because keen-eyed people may be able to make out its seven brightest stars. In all, the cluster probably contains in excess of 100 stars.

The Pleiades, also called M45, is a more concentrated cluster than the Hyades, and is much brighter, at a total brightness approaching 1st-magnitude. At a distance of almost 400 light-years, it is nearly three times as far away as the Hyades. Its stars are also much younger and hotter.

The brightest of the seven main Pleiades stars is Alcyone (magnitude 2.9). In descending order of brightness, the others are Atlas, Electra, Maia, Merope, Taygete, and Pleione.

THE CRAB

Close by Zeta, at the tip of the bull's southern horn, is another of Taurus's wonders. It is the Crab Nebula, or M1, the first in Charles Messier's list of clusters and nebulae. It is not an ordinary nebula, however. It is the gas cloud created when a star exploded as a supernova in A.D. 1054. Chinese astronomers recorded it as a "guest star." Today we call it a supernova remnant.

The Crab is just visible in powerful binoculars. Larger telescopes begin to resolve the delicate tapestry of glowing filaments in the gas cloud. Today, we can detect the remains of the star that exploded as a pulsar, which spins round 30 times a second.

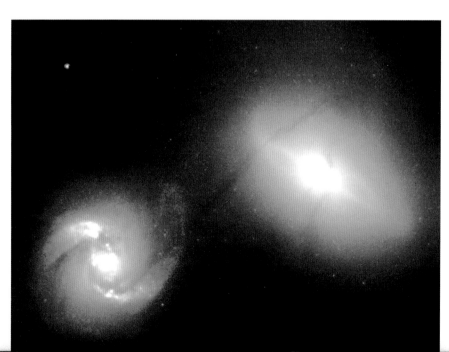

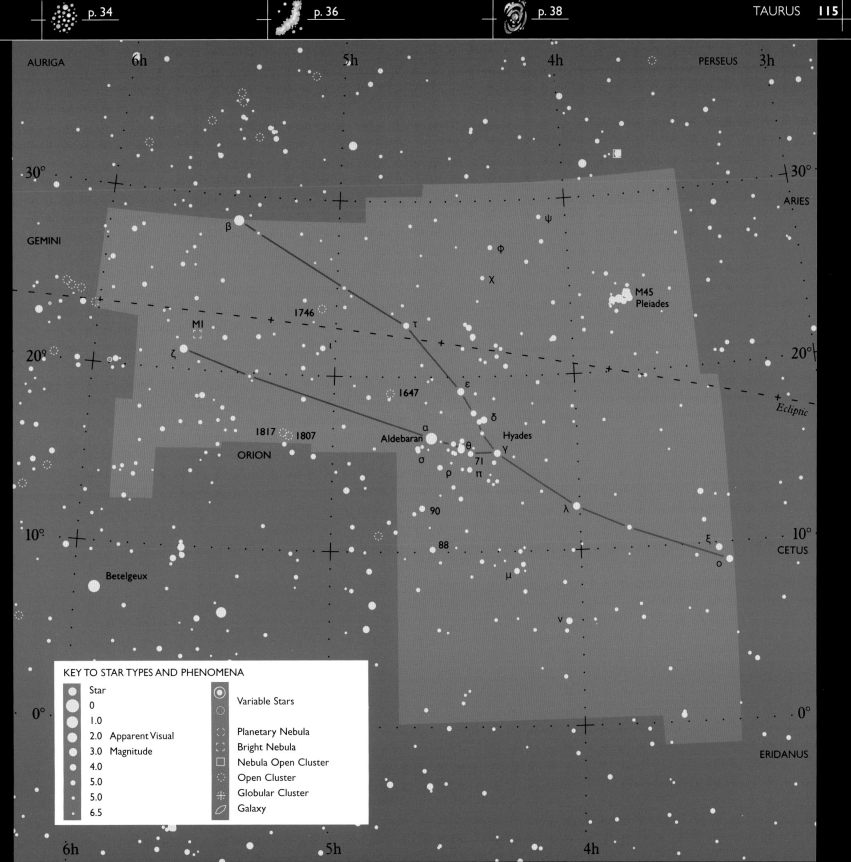

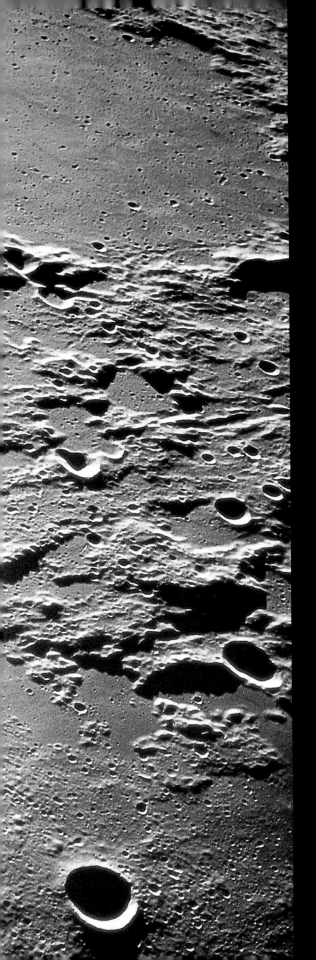

5 Sun and Moon

Two heavenly bodies dominate Earth's skies—the Sun by day and the Moon by night. During the day, the golden orb we know as the Sun arcs through the sky, pouring light and heat onto Earth, making it a bright, warm, and comfortable place for humans, animals, and plants to thrive in myriad different forms.

What is the Sun? It is a star, a searing hot globe of incandescent gas, just like the other stars in the night sky but much closer. Whereas the light from the stars takes years to reach us, the light from the Sun reaches us in only about eight minutes. Astronomers study the Sun closely, since its behavior not only critically affects conditions on Earth, but also mirrors the behavior of the majority of other stars in the Universe.

Just as the Sun travels across the sky by day, so the Moon travels across the sky by night. The Moon is our closest neighbor in space, right on our celestial doorstep. It is Earth's constant companion, its one and only natural satellite.

Coincidentally, both the Sun and the Moon appear about the same size in the sky. The Moon shines much less brightly because it merely reflects light from the Sun. This enables us to look at it safely with the naked eye and through binoculars and telescopes.

And what sights delight our eyes when we look at the Moon's surface! The photograph here hints at the pleasures that await us. There are flat plains, undulating hills, rugged highlands, wrinkles and ridges, and craters, always craters.

LEFT SOLAR REFLECTIONS
Sunlight picks out the features of the lunar landscape near the center of the Moon, throwing craters, hills, ridges, and rilles into sharp relief. Like all other bodies in the Solar System, we see the Moon only because it reflects light from the Sun.

Daytime Star

Our star of the day, the Sun, lies on average about 93 million miles (150 million km) away from Earth, a distance known as the astronomical unit (AU). As stars go, it is relatively small, with a diameter of about 865,000 miles (1,390,000 km). It gives off a yellowish light. Astronomers classify it as a yellow dwarf.

SURFACE AND ATMOSPHERE

The bright surface of the Sun is known as the photosphere (light-sphere). The temperature here is around 10,000°F (5,500°C). This compares with a temperature of about 30 million°F (15 million°C) at the Sun's center, where nuclear reactions produce the energy that keeps the Sun shining (see page 24).

Layers of thinner (less dense) gases surround the globe of the Sun, forming an atmosphere. The lower part of the atmosphere is known as the chromosphere (color-sphere) on account of its reddish hue. Above lies the Sun's outer atmosphere, called the corona.

Because the photosphere is so bright, we can't usually see the chromosphere or the corona. The only time we are able to is during a total solar eclipse, when the dark Moon blots out the Sun's bright face (see page 120).

THE STORMY SURFACE

The Sun's surface is a seething mass of boiling, bubbling gases and is in constant turmoil, like a stormy sea. Sometimes great fountains of fiery gas shoot out of the lower atmosphere and rise for hundreds of thousands of miles above the surface. Called prominences, they often form into loops, following the invisible lines of the Sun's powerful magnetic field.

Often, dark regions called sunspots appear on the surface. They are about 2,500°F (1,500°C) cooler than their surroundings. Triggered by magnetic effects, they come and go according to a solar, or sunspot, cycle of about eleven years. Sunspot Maximum marks the time of exceptional solar activity, when violent eruptions such as solar flares frequently take place.

THE SOLAR WIND

The Sun gives off a constant stream of electrically charged particles, which flow into space in all directions. We call this stream the solar wind. From time to time exceptional solar activity causes more particles than usual to stream out of the corona into space. We call these events coronal mass ejections.

At such times, the solar wind blows extra strong, and can affect Earth. It causes magnetic storms that can knock out power supplies and communications, and disable Earth-orbiting satellites.

This wind can also give rise to magnificent displays of the aurora in Earth's polar regions. These displays of shimmering curtains of colored light are named the aurora borealis (northern lights) in far northern skies and the aurora australis (southern lights) in far southern ones.

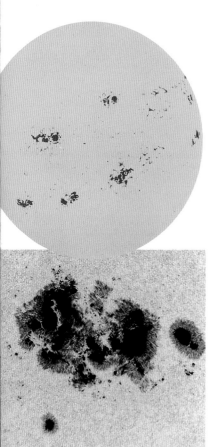

TOP AND ABOVE SUNSPOTS
Sites of exceptional magnetic activity on the solar surface pinpoint the location of sunspots (top). A well-developed sunspot group (above) can grow to more than 125,000 miles (200,000 km) across and persist for months. Typically, each sunspot has a dark center (umbra), surrounded by a lighter region (penumbra).

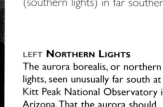

LEFT NORTHERN LIGHTS
The aurora borealis, or northern lights, seen unusually far south at Kitt Peak National Observatory in Arizona. That the aurora should appear at such latitudes indicates the occurrence of violent activity on the Sun, causing the solar wind to "blow" gale force.

DAYTIME STAR 119

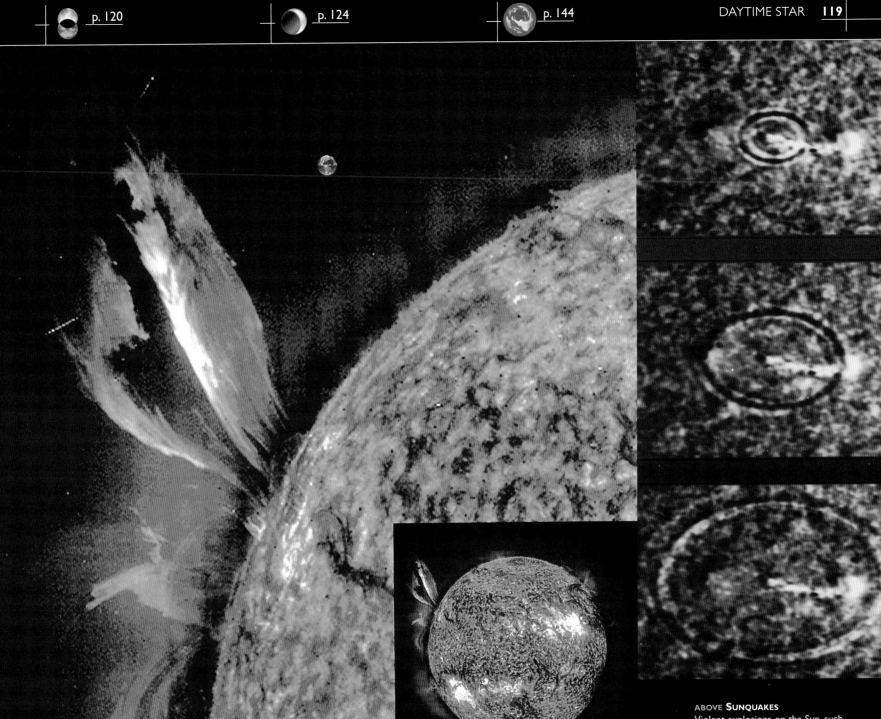

ABOVE STORMY SUN
The Sun's surface is seething as pockets of hot gases rise and fall, creating grainy patterns we call granulations. In this image from SOHO (Solar and Heliospheric Observatory) the brightest regions show the location of some of the most violent happenings on the Sun, gigantic explosions called solar flares. Shooting high above the surface are fountains of fiery gas known as prominences.

ABOVE SUNQUAKES
Violent explosions on the Sun, such as solar flares, can make the surface ripple—just as earthquakes can make Earth's surface ripple. For this reason, solar ripples like this are known as sunquakes.

Observing Solar Eclipses

Of all natural phenomena, none is as spectacular or as awe-inspiring as a total eclipse of the Sun. During a total eclipse, day turns suddenly into night, the air chills suddenly, and birds stop singing and start to roost. Astronomers travel all around the globe just to witness eclipses so that they can view the solar chromosphere and corona.

THE STRANGE COINCIDENCE

Eclipses of the Sun occur because of a strange astronomical coincidence. The Sun is 400 times farther away from us than the Moon and is 400 times bigger across than the Moon. This means that the Sun appears almost exactly the same size as the Moon in our skies.

From time to time, as the Moon travels in orbit around Earth, it lines up exactly with the Sun and Earth. When it lies on the opposite side of Earth from the Sun, it enters Earth's shadow, and we have an eclipse of the Moon, or lunar eclipse (see pages 124 and 125).

ABOVE BRILLIANT CORONA
Total eclipse viewed by the author on Bangka Island in Indonesia on March 18, 1988. The solar corona is visible as a brilliant pearly halo.

LEFT PINK PROMINENCES
Total eclipse viewed by the author in Hawaii on July 11, 1991, during one of the longest eclipses of the century. The chromosphere is visible around the dark limb of the Moon, and prominences leap high at 12 o'clock and 6 o'clock.

When the Moon comes between the Sun and Earth and is exactly lined up, it blots out the Sun's light and casts a shadow upon Earth. Then observers in the shadow see an eclipse of the Sun, or solar eclipse.

Partial and Total Eclipses

On average, solar eclipses occur about twice a year. Some are only partial eclipses, where only part of the Sun's disk is obscured by the Moon. A few are annular eclipses, where the Moon covers a large part of the Sun's disk, leaving only an annulus, or ring, around the edge visible. But the most exciting are total eclipses, where the Moon completely covers the Sun's disk and turns day suddenly into night, for a short time.

The Moon casts quite a small shadow in space, which only just touches Earth. The maximum size of complete shadow (umbra) is only about 170 miles (270 km) across. And, as the Moon moves, this shadow races across Earth's surface at a speed of several thousand miles an hour, sweeping out a "path of totality."

LEFT CLOUDED OUT
Astronomers may "chase" eclipses all around the world, but unfortunately they can't guarantee fine weather. The author hit unexpected cloud on the Zimbabwe border at high elevation in mid-summer during the total eclipse of December 4, 2002. The Sun was visible—just (top)—until a few minutes before totality. Then cloud obscured it completely. The twilight at totality, however, was still magical (bottom).

Totality—the period of total eclipse—can theoretically last up to about 7.5 minutes. Eclipses of over 6 minutes occur approximately every 18 years (1991, 2009, 2027, etc). But most other eclipses are considerably shorter—for example, on April 8, 2005, lasting just 42 seconds.

Eclipse Contacts

On average, the Moon takes about two hours to move across the face of the Sun. Astronomers describe the progress of an eclipse in terms of a series of "contacts."

First contact is when the Moon takes its first "bite" out of the Sun and marks the beginning of the partial phase of the eclipse. About an hour later, with the light fading perceptibly during this time, comes second contact, when totality begins. Just before it, comes the phenomenon of Baily's Beads, when beads of sunlight shine through the valleys on the limb of the Moon. During totality, the pink chromosphere becomes visible around the dark disk of the Moon. Then the Sun's pearly white corona appears. Third contact is when totality ends, and the Sun emerges once more, creating for an instant a beautiful "diamond-ring" effect. About an hour later the Moon completely clears the Sun. This is fourth contact, and the eclipse is over.

LEFT ORANGE HORIZON
During a total eclipse the sky does not go completely dark—this is due to light around the horizon, characteristically sunset-orange. Even to modern observers, this strange twilight seems eerie.

Queen of the Night

At an average distance of about 239,000 miles (384,000 km), the Moon is by far the closest heavenly body to Earth. The closest planet, Venus, never comes nearer to Earth than about 26 million miles (42 million km). The Moon is also special because it is the only other body in the Solar System yet visited by human beings.

BELOW **EARTHRISE**
Earth rises above the lunar horizon in this dramatic photograph taken by the Apollo astronauts. What a contrast there is between the drab and desolate surface of the Moon and colorful Earth far, far away.

A Double Planet

Compared with Earth, the Moon is a small body. Its diameter is only 2,160 miles (3,476 km), or about a quarter the size of Earth. This is actually extraordinarily large for a satellite in relation to the size of its parent planet. This is why astronomers sometimes consider Earth-Moon to be a double planetary system.

This view has further support in the way we think the Moon formed. Probably it was formed soon after Earth itself came into being, as a result of a collision between the infant Earth and another large body, maybe as big as Mars. When the collision occurred, a huge lump of matter was gouged out of Earth. This material, along with material from the other body, eventually came together under gravity to form the Moon.

Weak Gravity

Because it is so small, the Moon has much less mass than Earth. It therefore has a weak gravity, or "pull"—only one-sixth as strong as Earth's.

Due to its weak gravity, the Moon has been unable to hold on to any gases to make an atmosphere. This greatly affects conditions on the Moon. Without an atmosphere there is no weather, as we know it. And there are marked differences in temperature between lunar day and night, which each last about two Earth-weeks. In the lunar day temperatures soar as high as 250°F (120°C); while in the lunar night they can fall as low as -240°F (-150°C).

But, despite being weak, the Moon's gravity does have a significant affect on Earth. It tugs at the water in the oceans and creates the daily rise and fall of the tides.

Observing the Moon

Because it is so near and appears so large, the Moon is the ideal target for the astronomer with binoculars or a small telescope. These instruments reveal a wealth of detail that delights newcomers and old hands alike.

Even the naked eye reveals two distinct regions on the Moon—dark and light. The dark ones are the maria, or "seas." The light ones are the rugged highlands. The naked eye also reveals that the Moon always presents the same face toward us—we call it the near side. We can never see the far side from Earth.

The fascinating thing about lunar observing is that the appearance of the Moon changes nightly. We see more or less of its surface lit up as it goes through its phases during the month (see page 124). In particular, the boundary between light and shadow—lunar day and lunar night, moves across the surface.

We call this boundary the terminator. And it is on the terminator that the greatest detail can be seen. Because of the low Sun angle, long shadows bring craters, mountains, and other features into dramatic relief. Strangely perhaps, Full Moon is not the best time for detailed observations because the Sun is high and casts the least shadows.

QUEEN OF THE NIGHT

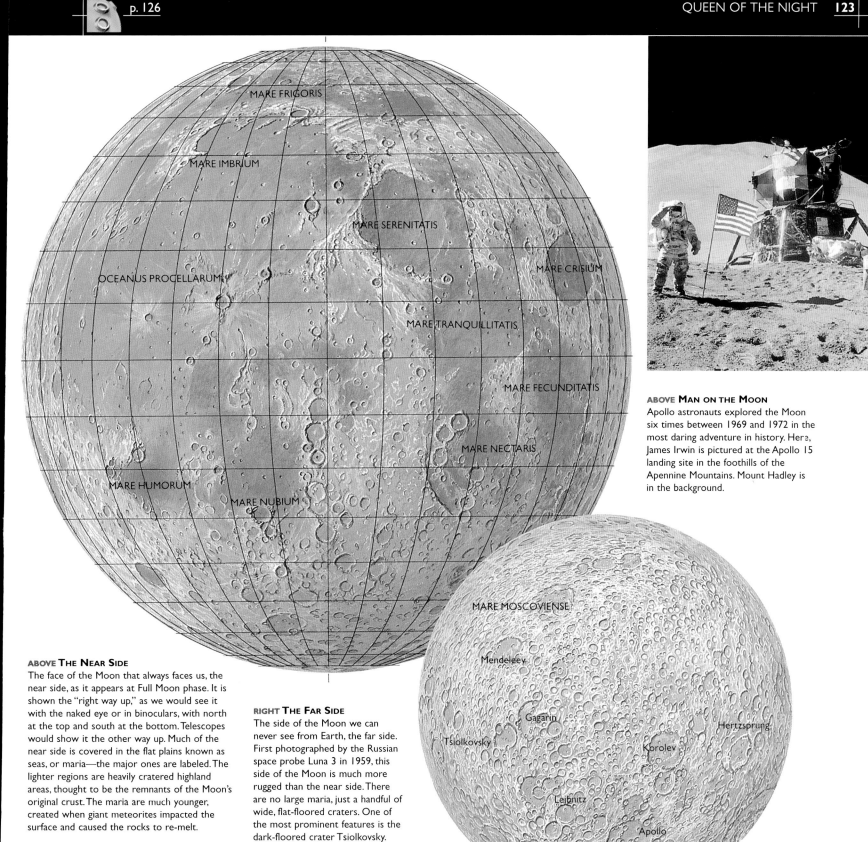

ABOVE MAN ON THE MOON
Apollo astronauts explored the Moon six times between 1969 and 1972 in the most daring adventure in history. Here, James Irwin is pictured at the Apollo 15 landing site in the foothills of the Apennine Mountains. Mount Hadley is in the background.

ABOVE THE NEAR SIDE
The face of the Moon that always faces us, the near side, as it appears at Full Moon phase. It is shown the "right way up," as we would see it with the naked eye or in binoculars, with north at the top and south at the bottom. Telescopes would show it the other way up. Much of the near side is covered in the flat plains known as seas, or maria—the major ones are labeled. The lighter regions are heavily cratered highland areas, thought to be the remnants of the Moon's original crust. The maria are much younger, created when giant meteorites impacted the surface and caused the rocks to re-melt.

RIGHT THE FAR SIDE
The side of the Moon we can never see from Earth, the far side. First photographed by the Russian space probe Luna 3 in 1959, this side of the Moon is much more rugged than the near side. There are no large maria, just a handful of wide, flat-floored craters. One of the most prominent features is the dark-floored crater Tsiolkovsky.

Moon Movements

As the Moon travels in orbit around Earth, we see more or less of its surface lit up by sunlight every night. We call these changing aspects of the Moon its phases. It takes the Moon 29.5 days to go through its phases. This time period is one of the great natural divisions of time, on which we base our calendar month.

OPPOSITE MOON IN SHADOW
During a lunar eclipse, the Moon enters Earth's shadow in space. The surface is lit up by sunlight refracted around Earth by the atmosphere, giving the Moon a distinctive reddish hue. A lunar eclipse can last as long as 2.5 hours.

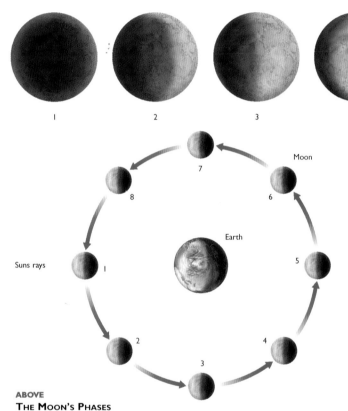

ABOVE
THE MOON'S PHASES
The different aspects, or phases, of the Moon as it circles Earth once a month (top). The diagram (above) helps explain how the phases come about. From Earth, we see more or less of the Moon lit up by sunlight as the Moon travels in its orbit. The numbers relate the positions of the Moon in its orbit at various times to the equivalent phase we view from Earth.

CHANGING PHASES

The phases begin with the New Moon. At this time we can't see the Moon at all because Sun, Moon, and Earth are in line, and the Sun is illuminating the Moon's far side. The near side is in shadow.

As the Moon moves along in its orbit, we see the Sun light up the edge of the Moon, and we see a slim crescent. Gradually, more and more of the nearside becomes visible. About seven days after New Moon, half of the nearside is lit up, and this is known as the First Quarter phase.

The illuminated area increases until, after about another seven days, the whole of the nearside is lit up. This is the Full Moon phase. Up to this point we say that the Moon has been waxing.

As the Moon moves on, less and less of the nearside is illuminated. We now say the Moon is waning. About a week after Full, only half of the nearside can be seen—this is the Last Quarter phase. After another week, only a slim crescent remains, until, 29.5 days after the New Moon, the nearside disappears at the next New Moon.

CAPTURED ROTATION

The Moon travels in orbit around Earth in space once every 27.3 days. And it also spins round once on its axis in the same time. This explains why the Moon always presents the same face—the near side—toward us. We say that the Moon has a captured rotation. Many other moons in the Solar System have captured rotation too.

RIGHT WAXING MOON
A waxing Moon about 10 days old. Interesting detail can be seen around Mare Imbrium. The bay Sinus Iridum is nicely outlined on the terminator. The flat floor of Plato is characteristically dark. Nearby, Alpine Valley runs through the Alps. Copernicus and its ray system are becoming prominent.

IN ECLIPSE

At certain times during the year, the Sun, Earth, and Moon line up exactly in space. This alignment causes eclipses to occur. When the Moon comes between the Sun and Earth, its shadow falls on part of Earth's surface, and a solar eclipse occurs (see page 120). This always happens at a New Moon.

When Earth moves exactly between the Sun and the Moon, the Moon moves into Earth's shadow and a lunar eclipse occurs. This always happens at a Full Moon.

Lunar Landscapes

The surface of the Moon is endlessly fascinating, dominated by the vast flat plains we call seas, or maria, and the more rugged highlands. Both types of landscape bear witness to the bombardment the Moon has suffered in the past by rocks from outer space, which have gouged out millions of craters, large and small.

BELOW LUNAR CRATERS
Craters pepper much of the Moon's surface. Found on the rugged far side of the Moon, this crater (IAU 308) measures about 50 miles (80 km) across. It is typical of large lunar craters, with sloping, terraced walls and a central mountain range on the floor.

The Maria

Maria cover the best part of the near side of the Moon, but are strangely absent on the far side. Largest by far is the sprawling Oceanus Procellarum, the Ocean of Storms. It merges in the north with the largest circular mare, Mare Imbrium, the Sea of Showers.

In the east, Mare Imbrium borders Mare Serenitatis, the Sea of Serenity, which merges into Mare Tranquillitatis, the Sea of Tranquility. Mare Tranquillitatis itself merges into Mare Nectaris, the Sea of Nectar, and Mare Fecunditatis, the Sea of Fertility.

The maria formed when huge space rocks smashed into the Moon's original crust. They gouged out huge basins, which were then filled with outpourings of molten lava. As a consequence, the maria are younger than the rest of the Moon's crust. Some are as young as 3.2 billion years old, around 1 billion years younger than the original crust.

The Craters

Some of the craters that pepper the lunar surface are the mouths of ancient volcanoes. But the majority were created by the impact of meteorites bombarding the surface from outer space. The classic lunar-impact crater has walls that rise about the surrounding landscape and a floor that dips below it.

One of the deepest lunar craters is the 70-mile (113-km) diameter Newton, close to the South Pole. Its walls rise over 1.2 miles (2 km) above the surrounding surface and plunge nearly 5.6 miles (9 km) below it.

Some of the larger craters, like Plato and Grimaldi, are notable for their low walls and flat, smooth floors. They are rather like miniature maria and are examples of walled plains.

When craters form, huge amounts of material are ejected from them. This "ejecta" is clearly visible around many of the younger lunar craters, such as Copernicus and Tycho. It fans out around the craters like brilliant "spokes" to form sparkling crater rays.

The floors of some of the craters near the Moon's poles are never touched by sunlight. And space probes such as Lunar Prospector have detected substantial amounts of iced water in them. This is good news for the future—if humans return to the Moon to set up permanent bases there.

Rilles And Ridges

As well as craters, volcanic activity on the Moon has produced several other features. The most prominent are called rilles, which are trench-like features that can be up to 3 miles (5 km) wide and a few hundred feet deep. Winding, or sinuous, rilles are probably ancient lava channels, marking where underground lava tubes have collapsed.

Just as there are sinuous depressions—rilles—so there are sinuous elevations, named wrinkle ridges. They snake across the surface of most mare regions.

LUNAR LANDSCAPES

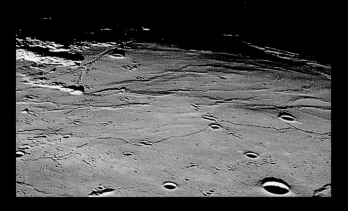

ABOVE
LUNAR PLAINS
The smooth surface of Mare Tranquillitatis typifies the flat plain regions on the near side of the Moon that are known as seas. There are only a few small craters.

RIGHT MOON ROCKS
Two kinds of rocks found widely on the Moon. Left is a kind of basalt, riddled with holes formed by gases escaping as the molten rock solidified. Right is a kind of breccia, a rock made up from bits of old rocks cemented together.

ABOVE IN THE ALPS
Cutting through the lunar Alps on the edge of Mare Imbrium is Alpine Valley, which runs for about 80 miles (130 km). The sinuous fault in the valley floor looks rather like a river bed.

RIGHT BIG BOULDER
On the Apollo 17 mission, geologist Harrison Schmitt inspects a giant boulder ejected by an ancient impact on the lunar surface. The location is a valley near the edge of Mare Serenitatis.

ABOVE **OCEAN LANDSCAPE**
This view of Oceanus Procellarum is centered on the crater Kepler, which becomes brilliantly lit at Full Moon. Away from the crater, the surface becomes smooth and flat, like most of Oceanus Procellarum.

The Moon Maps

The Moon maps that follow cover the lunar surface in four quarters, or quadrants. In the maps, the Moon appears "the right way up," as we view it with the naked eye or through binoculars. Through telescopes, of course, these views will appear upside-down.

The maps show the major lunar maria, or seas, together with prominent craters and mountains. The names of the maria are given in Latin, and the English equivalents can be found in the Glossary.

MOON MAP 1
THE NORTHWEST QUADRANT

Almost all of the northwest quadrant is occupied by maria. And just two dominate—Mare Imbrium and Oceanus Procellarum. With a diameter of around 700 miles (1,100 km), Mare Imbrium is the largest of the circular seas. It merges into Oceanus Procellarum, which sprawls over a vast area, making it by far the Moon's largest sea.

Mare Imbrium is well defined by arcs of high mountain ranges. In the south are the Carpathians, which peak at about 7,000 feet (2,100 m). Moving north are the Apennines, which run for some 280 miles (450 km). They contain some of the highest mountains on the Moon, which soar to more than 20,000 feet (6,000 m).

Farther north still, the Caucasus Mountains rise. Along with the highland region east of the Apennine, they form the boundary between Mare Imbrium and Mare Serenitatis.

Flanking Mare Imbrium in the north are the lunar Alps. Plato is the most prominent crater on the northern edge, noted for its

Key Features

Archimedes (47 miles, 75 km) has a flat, lava-filled floor.
Aristarchus (23 miles, 37 km) is small but gets very bright at Full Moon.
Copernicus (60 miles, 97 km) is a classic large lunar crater, with high-terraced walls and central mountain peaks. At Full Moon, it displays bright crater rays.
Eratosthenes (40 miles, 65 km) is a smaller version of Copernicus, located at the end of the Apennine mountain range.
Kepler (22 miles, 35 km) is small but stands out prominently at Full Moon because of its sparkling crater rays.
Lansberg (26 miles, 42 km), on the lunar equator, is a near-twin of Reinhold to the northeast.
Otto Struve (100 miles, 160 km) is one of the largest craters in this quadrant but is difficult to see because it lies near the western limb (edge).
Pico is an isolated mountain peak, rising to about 8,000 feet (2,400 m), near the edge of Mare Imbrium, just south of Plato.
Plato (60 miles, 97 km), on the edge of the Lunar Alps, is a circular crater noted for its flat, very dark floor.
Reinhold (30 miles, 48 km), south of Copernicus, has a deep floor.
Straight Range is a mountain range near the northern edge of Mare Imbrium. About 40 miles (60 km) long, it has peaks rising to 6,000 feet (1,800 m).

dark floor. More to the east and circled by the Jura Mountains, Sinus Iridum forms Mare Imbrium's most prominent bay. Measuring some 150 miles (250 km) across, it is a spectacular sight when the Moon is about 10 days old.

NORTHWEST QUADRANT

RIGHT IN THE NORTHWEST
Most of the quadrant is occupied by three seas—the sprawling Oceanus Procellarum, Mare Imbrium, and in the north, Mare Frigoris. The largest sea on the Moon, Oceanus Procellarum has no definite boundaries, gradually merging into other seas—like Mare Imbrium and Mare Humorum. Among the craters that grace this quadrant, Copernicus, Kepler, and Aristarchus are outstanding, being at the hub of spectacular crater-ray systems at Full Moon.

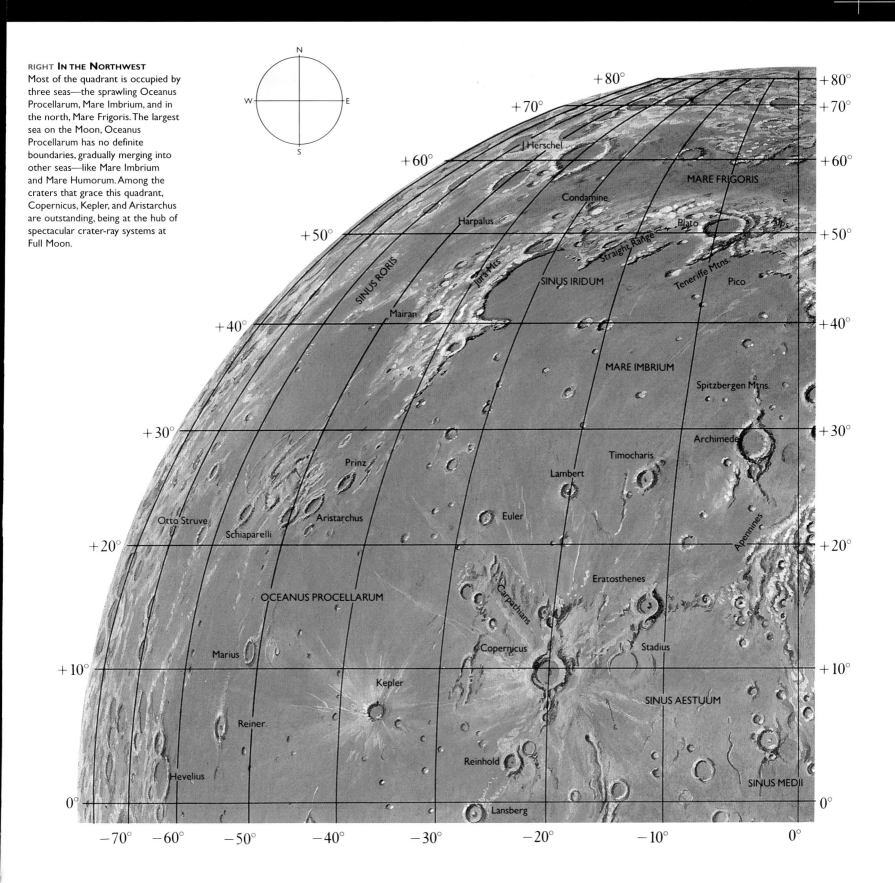

Moon Map 2
The Southwest Quadrant

The sprawling Oceanus Procellarum occupies much of this quadrant, covering an area of around 2 million square miles (5 million square km). This is over half as big again as the Mediterranean Sea on Earth. At its southern edge, Oceanus Procellarum merges into Mare Humorum and, through Mare Cognitum, into Mare Nubium.

Mare Humorum is one of the smaller circular seas, measuring about 280 miles (450 km) across. It is best seen when the Moon is about 11 days old. At its northern edge is the prominent crater Gassendi.

Mare Nubium is less well defined. Its most outstanding feature is a long cliff known as the Straight Wall. Some 800 feet (240 m) high, it is a fault in the Moon's crust that runs for about 100 miles (160 km).

Flanking Mare Nubium in the east is a spectacular chain of large craters going south from the large Ptolemaeus. The chain looks stunning at First Quarter, when the terminator (the boundary between light and shadow) shows them up in bold relief.

Among the other craters that grace this quadrant, Tycho is outstanding because it is the center of the most extensive crater-ray system on the Moon. The rays extend away from the crater for thousands of miles.

Two of the six Apollo landings were made in this quadrant a few degrees south of the lunar equator: Apollo 12 set down on Oceanus Procellarum; Apollo 14 near the Fra Mauro formation.

Key Features

Alphonsus (80 miles, 129 km) is the largest of three walled plains on the 0° longitude line. Rilles and mountains cross its floor. It is best seen at First Quarter phase.

Bailly (183 miles, 295 km) is the Moon's largest crater, but it is rarely seen because it lies right on the southern limb.

Bullialdus (31 miles, 50 km), on the edge of Mare Nubium, is a perfectly formed crater, with intact walls and central peaks.

Clavius (144 miles, 232 km) is the largest crater we can see well from Earth. Its walls and floor are peppered with craters.

Gassendi (55 miles, 89 km) borders Mare Humorum. It is a fine-walled plain.

Grimaldi (120 miles, 193 km) lies near the limb just south of the Equator. It is noted for its dark floor.

Longomontanus (90 miles, 145 km) is one of a trio of large craters in the south, along with Clavius and Maginus. Like them, it has partly ruined walls.

Maginus (110 miles, 177 km) lies east of Longomontanus. These two craters, along with Tycho in the north and Clavius in the south, make up a kind of lunar "southern cross."

Pitatus (50 miles, 80 km), at the southern edge of Mare Nubium, has a dark floor with a low central peak.

Ptolemaeus (92 miles, 148 km) is the first and largest of a chain of large craters that extend north-south near the 0° longitude line. It is a walled plain with a dark floor.

Riccioli (100 miles, 160 km) lies just south of the equator near the limb. Like its neighbor Grimaldi, it has a dark floor.

Schickard (124 miles, 200 km), near the southern limb, is a fine-walled plain, with low walls.

Tycho (52 miles, 84 km) vies with Copernicus for being the finest lunar crater. At Full Moon it becomes the center of the most brilliant crater-ray system.

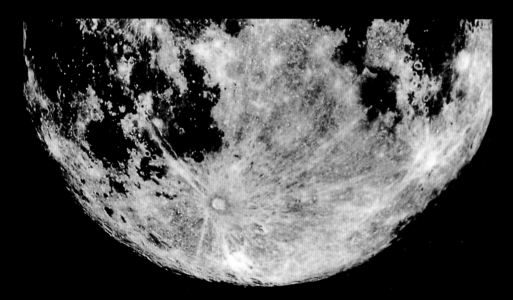

ABOVE TYCHO'S RAYS
At Full Moon, Tycho outshines nearly every other feature on the Moon. Its brilliant crater rays extend in all directions, as far as Oceanus Procellarum in the northwest and into Mare Nectaris in the southeast.

TOP TYCHO CRATER
Close-up photograph of the southern crater Tycho, taken by a U.S. Lunar Orbiter probe from an altitude of about 135 miles (220 km). It shows the terraced walls and central mountain peaks typical of large lunar craters.

SOUTHWEST QUADRANT

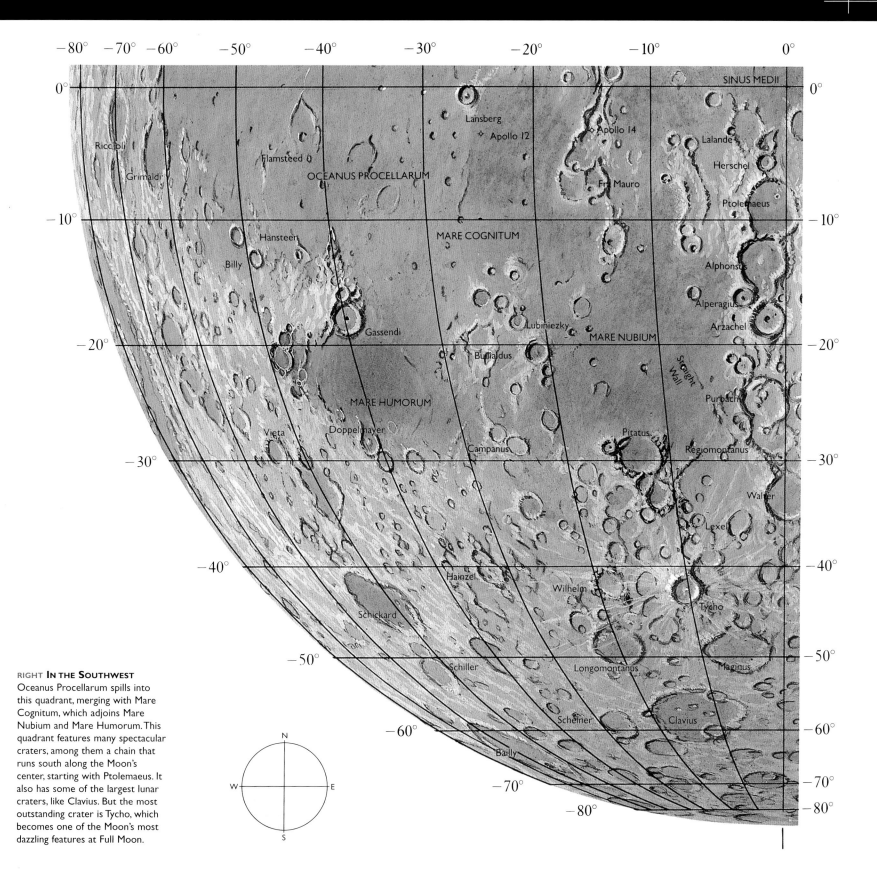

RIGHT IN THE SOUTHWEST
Oceanus Procellarum spills into this quadrant, merging with Mare Cognitum, which adjoins Mare Nubium and Mare Humorum. This quadrant features many spectacular craters, among them a chain that runs south along the Moon's center, starting with Ptolemaeus. It also has some of the largest lunar craters, like Clavius. But the most outstanding crater is Tycho, which becomes one of the Moon's most dazzling features at Full Moon.

Moon Map 3
The Northeast Quadrant

Two of the largest seas that occupy much of this quadrant are Mare Serenitatis and Mare Tranquillitatis. They are roughly the same size, about 500 miles (800 km) across. To the east, toward the limb, is Mare Crisium, which measures about 300 miles (500 km) across. Distinctly circular, it is well defined by prominent highlands.

Mare Serenitatis is bordered in the west by the Apennines and the Caucasus Mountains, which have peaks soaring over 20,000 feet (6,000 m). These ranges make a boundary with Mare Imbrium.

Forming the southern border of this mare are the rather lower Haemus Mountains. They give way in the east to an area of low hills and small craters, the largest of which is Plinius. The surface of the mare is noticeably wrinkled but has only one significant crater, Bessel.

In the north, Mare Serenitatis merges, through Lacus Mortis and Lacus Somniorum, into Mare Frigoris. This irregular, elongated sea stretches west and eventually meets up with Oceanus Procellarum.

In the south, Mare Serenitatis peters out in a hilly region and merges into Mare Tranquillitatis. This sea in turn merges in the southeast near the lunar equator into Mare Fecunditatis.

Mare Crisium, near the eastern limb, is the first sea revealed as the Moon goes through its phases, being immediately identified at the new crescent. The other mare in this quadrant, the small Mare Vaporum, is best seen at First Quarter, being located close to the Moon's centerline.

Three of the six Apollo landings took place in this quadrant. Apollo 11 landed in the south of Mare Tranquillitatis; Apollo 15 in the Apennines; and Apollo 17 near Littrow crater on the edge of Mare Serenitatis.

Key Features

Alpine Valley cuts through the lunar Alps, in effect linking Mare Imbrium and Mare Frigoris. Although it may look like a river valley, it is actually a particularly straight geological fault.

Aristillus (35 miles, 56 km) is one of a pair (with Autolycus) of small but prominent craters near the eastern edge of Mare Imbrium. It has a deep floor.

Aristoteles (60 miles, 97 km) is one of a pair (with Eudoxus) of craters north of Mare Serenitatis.

Atlas (55 miles, 89 km) forms one of another pair (with Hercules) of craters in the north.

Autolycus (22 miles, 36 km) forms a conspicuous pair of craters with Aristillus.

Bessel (12 miles, 19 km) is a tiny crater, but the largest in Mare Serenitatis.

Cleomedes (78 miles, 126 km) is a large crater just north of Mare Crisium and looks wonderful just after Full Moon.

Eudoxus (40 miles, 64 km) forms a prominent pair of craters with Aristoteles.

Gauss (85 miles, 136 km) is quite large, but difficult to see because it lies near the limb.

Hercules (45 miles, 72 km) forms a pair of craters with Atlas.

Hyginus Rille, one of the Moon's longest rilles, runs from the south of Mare Vaporum toward Mare Tranquillitatis.

Manilius (22 miles, 36 km), on the edge of Mare Vaporum, has reflective walls that make it one of the quadrant's brightest spots.

Menelaus (20 miles, 32 km) is another bright crater, on the opposite side of the Haemus Mountains from Manilius.

Posidonius (60 miles, 96 km), on the edge of Mare Serenitatis, is a beautiful crater with interesting walls.

ABOVE MOONWALKER Apollo 11 astronaut Edwin "Buzz" Aldrin, pictured in a famous photograph taken on Mare Tranquillitatis at the site of the first lunar landing in July 1969. Pictured in his vizor is first-man-on-the-Moon Neil Armstrong and the Apollo 11 lunar landing module "Eagle."

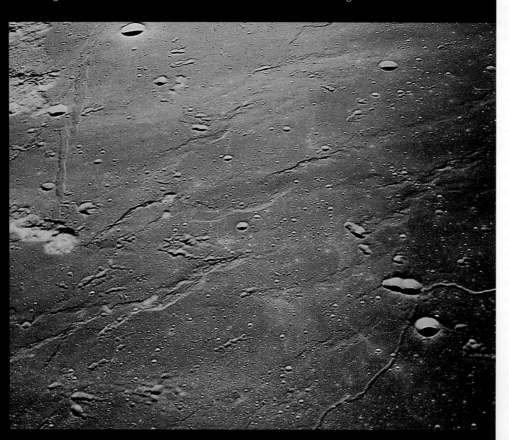

RIGHT A breathtaking view of the southwest region of Mare Tranquillitatis, close to the site of the first lunar landing. The Apollo 11 lunar module set down near the top center of the photograph. In the foreground is the sinuous Maskelyne Rille.

NORTHEAST QUADRANT 133

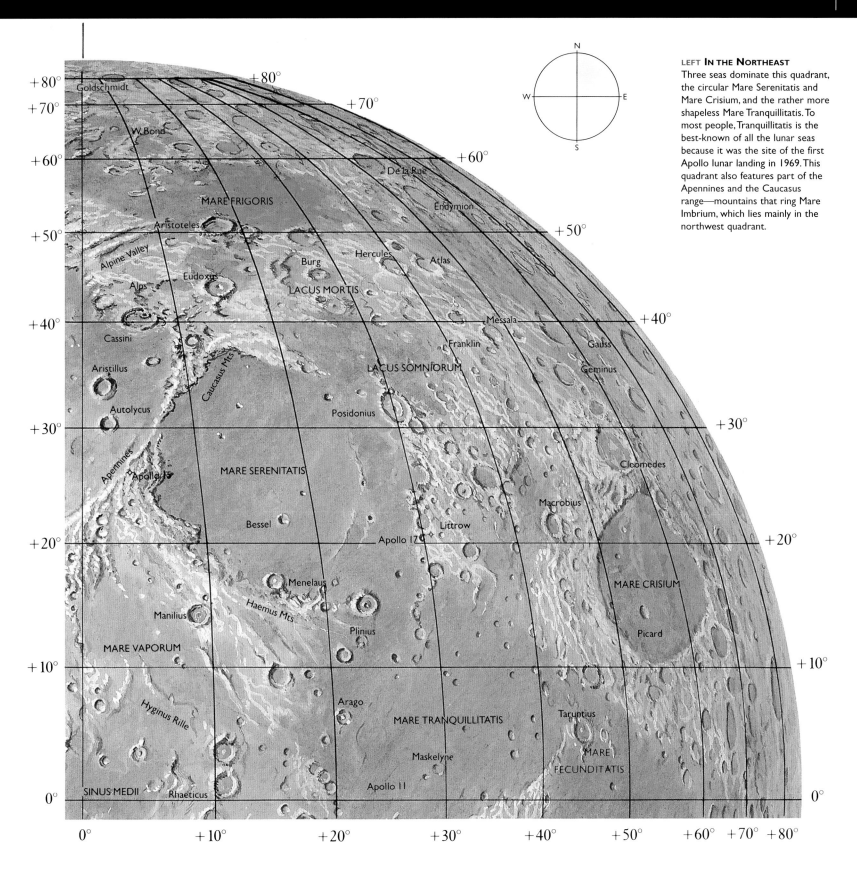

LEFT IN THE NORTHEAST
Three seas dominate this quadrant, the circular Mare Serenitatis and Mare Crisium, and the rather more shapeless Mare Tranquillitatis. To most people, Tranquillitatis is the best-known of all the lunar seas because it was the site of the first Apollo lunar landing in 1969. This quadrant also features part of the Apennines and the Caucasus range—mountains that ring Mare Imbrium, which lies mainly in the northwest quadrant.

Moon Map 4
The Southeast Quadrant

Generally more rugged than the other quadrants, this quadrant has two main mare regions, Mare Nectaris and Mare Fecunditatis. Mare Nectaris is roughly circular and measures about 300 miles (500 km) across. Straddling the lunar equator, Mare Fecunditatis is more irregular in shape and is about twice as wide.

Mare Nectaris is quite well defined, flanked in the west by a trio of large craters, Theophilus, Cyrillus, and Catharina. Beyond this arc of craters lies the only prominent mountain range in this quadrant, the Altai Mountains, or Scarp.

The Altai range runs for about 300 miles (500 km) from near Catharina to Piccolomini, and peaks reach a height of about 13,000 feet (4,000 m). The range was probably formed by the same impact that created Mare Nectaris itself.

An area of cratered highlands separates this sea from the less well-defined Mare Fecunditatis. The eastern edge of the sea is bounded by a chain of large craters, including Langrenus and Vendelinus.

Beyond Mare Fecunditatis, right on the limb and straddling the lunar equator, is the smaller Mare Smythii. It can be seen best shortly after Full Moon. On the limb in the far south, Mare Australe can also be glimpsed at about this time.

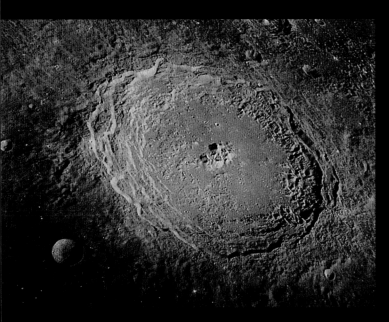

ABOVE THE QUADRANT'S FINEST
The crater Langrenus, photographed from orbit by Apollo astronauts. It has classic high terraced walls and central mountain peaks. It is conspicuously bright most of the time and is the most prominent of the chain of craters near the limb in this quadrant.

BELOW MOON ROCK
This was part of the haul of Moon rocks brought back by the Apollo 16 astronauts, who landed in a highland region near Descartes crater. Like all lunar rocks, it is volcanic.

Key Features

Albategnius (80 miles, 129 km) is an ancient walled plain east of Ptolemaeus. It has a deep crater, Klein, embedded in its walls.

Aliacensis (52 miles, 84 km) is the largest of a chain of craters in the west, close to and parallel with the crater chain on the 0° longitude line.

Fracastorius (60 miles, 97 km) is a badly ruined crater almost obliterated during the formation of Mare Nectaris, of which it now forms a bay.

Hipparchus (90 miles, 145 km) is an ancient walled plain close to the Moon's center. It forms a pair with Albategnius.

Langrenus (85 miles, 137 km) is the brightest of the string of large craters that more or less follow the 60° longitude line near the northeastern limb.

Maurolycus (68 miles, 109 km), west of Stofler, is an ancient, badly eroded crater.

Petavius (106 miles, 170 km) is one of the large craters on the 60° longitude line. It displays beautiful crater rays at Full Moon, which meet those coming from Tycho.

Piccolomini (50 miles, 80 km) lies at the southern end of the Altai mountain range.

Rheita (42 miles, 68 km) lies next to one of the finest crater chains on the Moon, the 100-mile (160-km) Rheita Valley.

Theophilus (62 miles, 100 km) is the largest of the arc of craters at the western edge of Mare Nectaris. It has well-terraced walls and a deep floor with central peaks.

Vendelinus (103 miles, 165 km) lies roughly between Langrenus and Petavius, but is not as prominent.

Walter (80 miles, 129 km) is the southernmost of the string of large craters that lie on the 0° longitude line.

SOUTHEAST QUADRANT

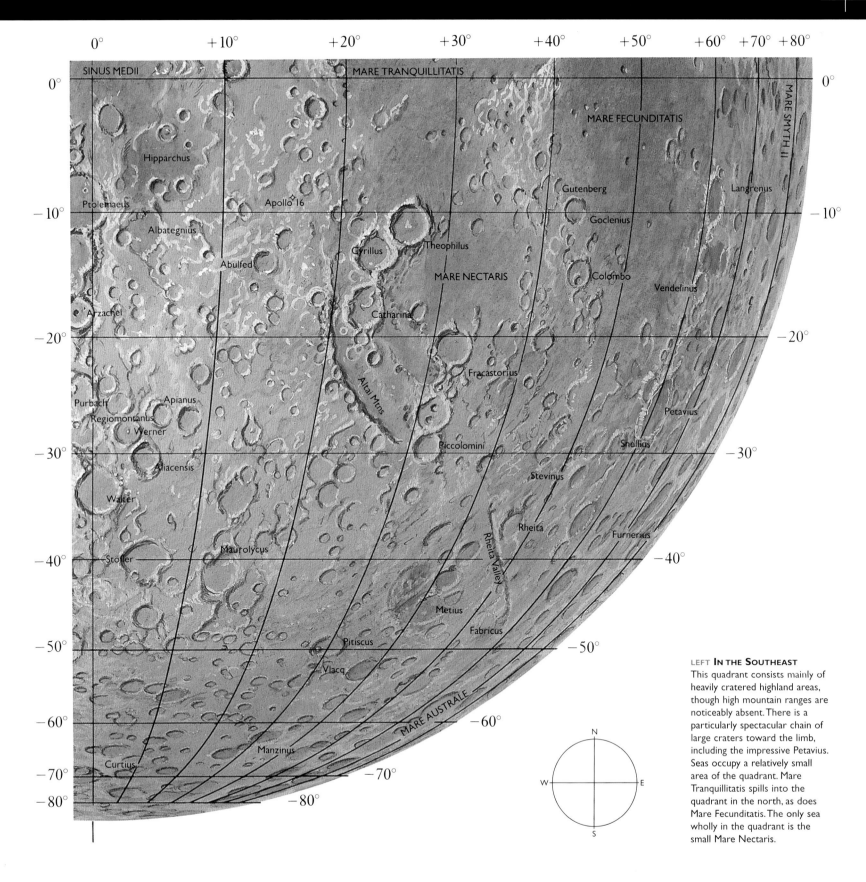

LEFT IN THE SOUTHEAST
This quadrant consists mainly of heavily cratered highland areas, though high mountain ranges are noticeably absent. There is a particularly spectacular chain of large craters toward the limb, including the impressive Petavius. Seas occupy a relatively small area of the quadrant. Mare Tranquillitatis spills into the quadrant in the north, as does Mare Fecunditatis. The only sea wholly in the quadrant is the small Mare Nectaris.

6 The Sun's family

The Sun was born about five billion years ago in one of the cold, dark molecular clouds that are found in interstellar space. When it was born, it was surrounded by a flattened disk of gas and dust. Over a period of a half-billion years, this disk condensed and fragmented into the motley collection of bodies that now form part of the Sun's family, or Solar System. With 750 times the mass of all these other bodies combined, the Sun holds them together with its powerful gravity, which reaches out halfway to the nearest stars.

Earth is one of the larger bodies in the Solar System that we refer to as planets. There are eight other planets—four smaller than Earth, and four very much larger. We can see five of the other planets—Mercury, Venus, Mars, Jupiter, and Saturn—with the naked eye, wandering across the heavens.

In addition to the planets, the Sun's family also includes more than 135 moons, thousands of the miniplanets we call asteroids, and countless numbers of icy lumps that become visible as comets when they travel in toward the Sun. Even specks of interplanetary dust reveal their presence in the night sky when they burn up in the atmosphere as meteors, or shooting stars.

All these heavenly bodies make rewarding study with the naked eye, as well as through binoculars and telescopes. But most information about them has come from the pictures and data returned by space probes. Probes have now visited all the planets except Pluto.

LEFT ASPECTS OF SATURN
Saturn, the undisputed jewel of the Solar System. We see different aspects of Saturn's glorious rings as the planet travels along its 30-year orbit around the Sun. These views, taken by the Hubble Space Telescope, show the rings opening up between 1996 (bottom) and 2000 (top).

The Solar System

Until about 500 years ago, people believed that Earth was the center of the Universe, and that all the other heavenly bodies circled round it. This idea, elaborated by the Greek astronomer Ptolemy in about A.D. 150, is known as the Ptolemaic system.

BELOW THE CIRCLING PLANETS
The planets in their orbits around the Sun, drawn approximately to scale. The four inner planets are relatively close together, but the five outer planets are widely separated. Viewed from the "north" of the Solar System, the planets travel around the Sun in a counterclockwise direction. All the planets except Venus also spin counterclockwise on their own axis.

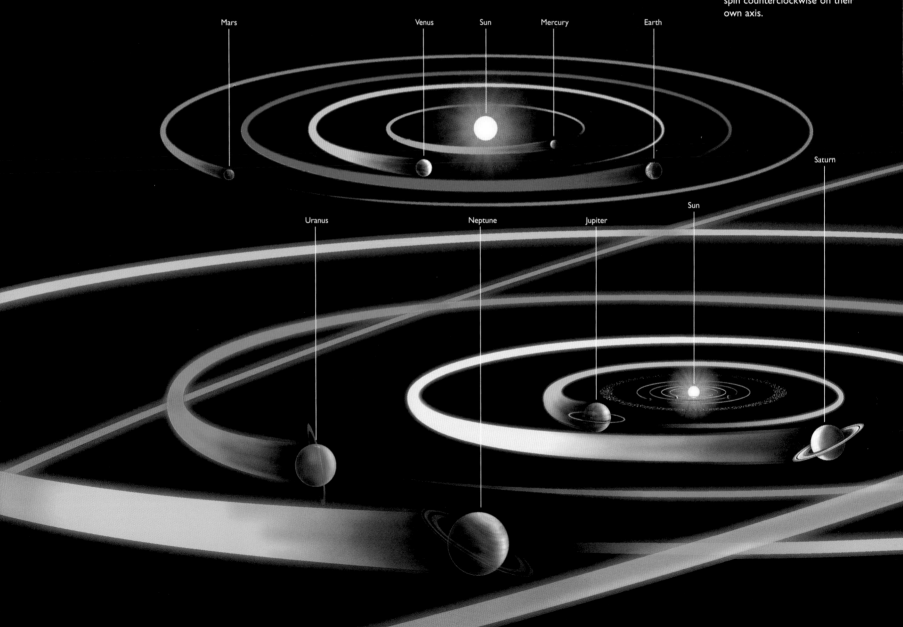

Sun-Centered

The Ptolemaic system, however, could not adequately explain the motion of the planets, in particular Mars, which appears to back-track in the heavens at times.

But, in 1543, Copernicus put matters straight, realizing that planetary motions could better be explained if the Sun, and not Earth, were at the center of the Universe. Because he relegated Earth to the status of a mere planet, he incurred the displeasure of the Church, which regarded his ideas as heresy. The Copernican system, therefore, was not generally accepted until the next century, after Johann Kepler had established the definitive laws of planetary motion.

In Copernicus's day, only six planets were known—Mercury, Venus, Earth, Mars, Jupiter, and Saturn. Three more have since been discovered—Uranus (in 1781), Neptune (in 1846), and Pluto (in 1930).

Pluto sometimes wanders over 4.5 billion miles (7 billion km) from the Sun. But the Solar System extends very much farther than this. New distant bodies are being found all the time, such as the "planetoid" Sedna (in 2004), which orbits 8 billion miles (13 billion km) from the Sun.

Nicolaus Copernicus (1473–1543)

Copernicus was a Polish priest-astronomer who laid the foundations of modern astronomy. He put our little corner of the Universe in order by advancing the theory of the Solar System, in which Earth and the other planets circle around the Sun. He put forward his ideas in a book *Concerning the Revolutions of the Celestial Spheres* which was published as he lay on his deathbed.

An illustration from his book (below) shows the six planets known at the time circling around a stationary Sun.

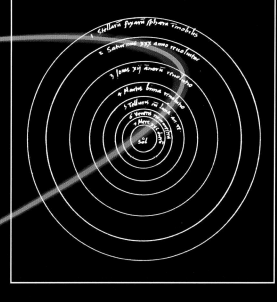

LEFT ECCENTRIC PLUTO Pluto's orbit takes it way above and way below the general orbital plane of the other planets. The orbit is also highly eccentric, taking Pluto sometimes within the orbit of Neptune. It traveled inside Neptune's orbit between 1979 and 1999, but it has now regained its rightful position as the most distant planet.

Observing the Planets

Ancient astronomers were familiar with Mercury, Venus, Mars, Jupiter, and Saturn because they are visible to the naked eye. They named these bright "stars" planets, meaning "wanderers" because they wander around the constellations. The other, so-called "fixed stars" always stay in the same position in the constellations.

OPPOSITE **JUPITER'S BANDS**
Through a telescope, dark and light bands can readily be distinguished in the atmosphere of Jupiter. But the most prominent feature is the Great Red Spot.

Bright Spots

The five naked-eye planets are always worth observing because they are constantly changing—their brightness and their position in the sky.

It is interesting, for example, to see how they change the appearance of the constellations they pass through. They always travel through the constellations of the zodiac, like the Sun. This is because they all travel through space on more or less the same plane as the Sun.

Because of their brilliance, Venus, Jupiter, and often Mars, are easy to spot in the night sky. Mercury and Saturn are more elusive. But you can find out exactly where the planets are at any time by consulting astronomical magazines and yearbooks.

Looping the Loop

The movements of the planets through the heavens are not entirely straightforward. For example, Mars, Jupiter, and Saturn, which usually travel eastward against the background of stars, sometimes start traveling westward.

This retrograde motion happens because Earth, traveling more quickly inside their orbits, periodically catches up with and then overtakes them. And they appear to "loop the loop" in the sky.

RIGHT **MARS'S MARKINGS**
Telescopes show a variety of features on Mars. Prominent in this view of the planet is the north polar ice cap. Vague dark markings can be seen in other regions, which change with the seasons.

Mercury

Of the five naked-eye planets, Mercury is the most difficult to spot, even though it lies relatively close to us. The reason is because the planet always stays close to the Sun.

Under favorable conditions, Mercury may sometimes be seen low down near the horizon in the west just after sunset as an "evening star." At other times it may be seen just before dawn in the east as a "morning star." At maximum brightness, it reaches about magnitude -1.5.

Venus

By contrast, Venus is by far the easiest planet to recognize. It is the evening star that shines brightly quite high in the sky in the west after sunset on many nights of the year. It is brilliant, reaching a maximum magnitude of about -4.7, far outshining all the stars and all the other planets. It can also be a brilliant morning star, seen in the east at dawn.

Small telescopes will be able to spot that Venus shows phases, like the Moon does.

Mars

Mars, when bright, is unmistakable in the night sky because of its pronounced reddish hue; hence its common name, the Red Planet.

Mars varies markedly in brightness according to where it is in its orbit. When it is at opposition—and closest to Earth—it can become as bright as magnitude -2.8, rivaling Jupiter in brilliance.

OPPOSITE **SATURN'S RINGS**
Through a telescope, Saturn's A (outer) and B rings, and the Cassini Division between them, show up clearly. The faint inner C ring is often difficult to make out.

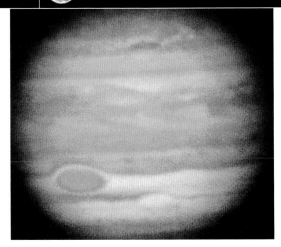

Jupiter

Although Jupiter is much farther away than Mars, it is also much bigger. At its most brilliant it can outshine the Red Planet, reaching a magnitude of -2.9. The two planets can easily be distinguished because Mars is reddish, while Jupiter is brilliant white.

Saturn

Saturn is a giant-sized planet like Jupiter. But it lies twice as far away, and so never becomes as bright. Indeed, at maximum brightness, Saturn is the faintest of the naked-eye planets, reaching a maximum magnitude of only about -0.3.

Nevertheless, at peak brightness, it outshines most of the stars and is easy to follow. At other times it tends to get lost among the other stars.

ABOVE PROBING THE PLANETS
We have learned the most about planets from space probes, which have traveled deep into interplanetary space to spy on them from close quarters. Outstanding have been two Voyager probes of this design, which set out in 1977 to explore the outer planets. Their instruments included cameras, radiometers, and particle detectors. They relayed information back to Earth from their 12-foot (3.7-m) diameter antenna.

ABOVE VOYAGER'S SATURN
Saturn viewed in close up by Voyager 2 in 1981. False colors have been used to bring out the greatest possible detail in the cloud bands in the atmosphere and in the ring system.

The Scorching Planets

Mercury and Venus are rocky planets like Earth and are oven-hot. But otherwise they are quite different. Mercury has scarcely a trace of an atmosphere and has a heavily cratered, lunar-like surface, while Venus has a thick, torrid atmosphere and has been shaped by volcanoes.

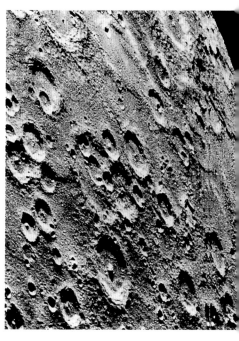

ABOVE **Craters Galore**
A Mariner close-up of Mercury's surface, showing the extensive cratering that covers the planet. The larger craters resemble those on the Moon, with terraced walls and central mountain peaks. Numerous fault lines are also visible in the picture.

What Mercury is Like

Mercury is the second smallest planet, after Pluto. With a diameter of 3,031 miles (4,878 km), it is about half as big again as the Moon. And, like the Moon, its mass and gravity are too low for it to have retained any appreciable atmosphere.

With nothing to protect it from bombardment by space rocks, it is almost completely covered with craters. This makes the surface look similar to the heavily cratered areas of the Moon. There are a few small plain regions, but nothing like the vast maria, or "seas," found on the Moon.

One particularly large space rock created the biggest feature we know on Mercury—the huge Caloris Basin. Some 870 miles (1,400 km) across, it is ringed by mountain chains.

ABOVE **Mariner's Mercury**
A mosaic of about 20 images of Mercury taken by the Mariner 10 probe as it flew past the planet in March 1974. Mariner made two further passes six months later and in March 1975. But altogether it imaged only about half of Mercury's surface. The Mercury Messenger probe will map the whole planet when it goes into orbit around it in 2009.

Mercury Motions

Mercury is the planet closest to the Sun and therefore would be expected to be hot. But it gets superhot—with temperatures peaking at 800 °F (430°C) or above—because of its slow rate of rotation.

Mercury takes 59 days to spin around once on its axis (compared with Earth's 24 hours). This means that, as the planet travels in its 88-day orbit around the Sun, a point on the surface is exposed to sunlight for 176 days at a time. And it is this that makes it oven hot. That point then spends 176 days in darkness, which allows the temperature to drop as low as -290°F (-180°C).

VENUS, PLANET FROM HELL

With a diameter of 7,521 miles (12,104 km), Venus is a near-twin of Earth in size, and it also has an atmosphere. But the two planets couldn't be more different otherwise.

Whereas Earth is a green and pleasant land, Venus is a hellish place. The atmosphere is made up mainly of suffocating carbon dioxide and is so thick that its pressure is 90 times the atmospheric pressure on Earth.

The dense carbon dioxide atmosphere acts like a greenhouse to trap the Sun's energy, creating a runaway "greenhouse effect." This has led to global warming that pushes temperatures on Venus to as high as 900°F (480°C).

VENUS LANDSCAPES

Thick sulfuric clouds in Venus's atmosphere completely hide the planet's surface from view. But we now know what it is like because orbiting space probes like Magellan have used radar to "see" through the clouds.

The whole landscape of the planet has been shaped by volcanoes. Thousands of them are scattered over the globe, and outpourings of lava from them have created a landscape of extensive rolling volcanic plains.

Most of the surface is low-lying, with only two main upland regions, or "continents." They are Aphrodite Terra near the equator, which is about the size of Africa, and the rather smaller Ishtar Terra in the north.

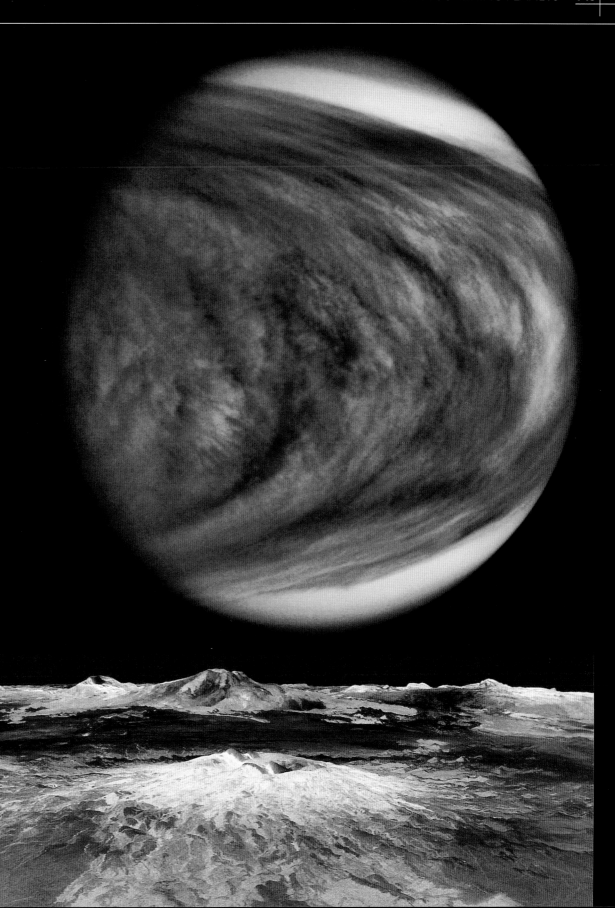

ABOVE RIGHT VENUS'S ATMOSPHERE
In 1978, the Pioneer Venus probe pictured Venus in ultraviolet light and was able to pick out chevron-like cloud patterns in the atmosphere. The clouds are not like those on Earth, but are made up of droplets of corrosive sulfuric acid.

RIGHT VENUS'S VOLCANOES
Maat Mons is one of the many volcanoes that have shaped the surface of Venus. Some 4 miles (6 km) high, it has erupted repeatedly over millions of years, spewing lava over the surrounding area. The paler areas in the picture show the most recent lava flows.

The Red Planet

Mars is well named the Red Planet because of its distinctive reddish appearance in the sky. It comes closer to Earth than any other planet except Venus, sometimes approaching as close as 35 million miles (56 million km). And it is the planet that most closely resembles Earth.

BELOW **HUBBLE'S MARS**
Mars at opposition in June 2001, when the planet was less than 43 million miles (68 million km) away. This Hubble Space Telescope picture shows dust storms sweeping the planet. They are noticeable over the north polar ice cap (top) and around the Hellas region (lower right).

Earth-like

Although Mars (diameter 4,222 miles, 6,794 km) is only about half the size of Earth, it resembles our own planet in several ways. It spins around once on its axis in 24 hours 37 minutes, and so its "day" (named a "sol") is only fractionally longer than our own day. Its axis is tilted at about 25 degrees, close to Earth's 23½ degrees. This axial tilt means that Mars has seasons like Earth, though they are nearly twice as long because of Mars's longer "year" (687 Earth-days).

There are other similarities too. Mars has an atmosphere like Earth, though it is very much thinner, with only about one-hundredth the pressure of Earth's atmosphere, and the main gas present is carbon dioxide. But clouds float in the atmosphere, frost coats the landscape in winter, and ice builds up to form brilliant white caps at the poles.

Martian Life

It was because of such similarities that people once believed that Mars might harbor life, even intelligent life. However, investigations by space probes, which have showed Mars to be a cold (with temperatures down to -190°F, -125°C), barren, and inhospitable world, have all but ruled out this possibility. But the latest findings from probes, such as the Pathfinder lander, the Mars Odyssey, Mars Global Surveyor, and Mars Express orbiters, and the twin rovers Spirit and Opportunity, suggest that Mars once had a milder climate and was awash with water. It is possible that life might have evolved in those conditions, only to perish when the climate deteriorated.

LEFT **THE THARSIS VOLCANOES**
This Viking approach picture shows Mars's four outstanding volcanoes. In a line near the center are the three volcanoes on the Tharsis Ridge. From left to right, they are Arsia Mons, Pavonis Mons, and Ascreus Mons. Above them is the massive crater of the volcano Olympus Mons.

Martian Landscapes

Mars is a world of great contrasts. Much of the southern hemisphere is rugged, heavily cratered terrain. There are two huge impact basins—Hellas, more than 1,000 miles (1,600 km) across, and Argyre, about half the size. By contrast, the northern hemisphere is dominated by vast deserts, which astronomers think might once have been oceans.

LEFT ON THE SURFACE
Much of the surface of Mars is littered with chunks of rocks. The Pathfinder probe returned this image of the surface in Ares Vallis in 1997. It is thought to be the estuary of an ancient river system, and the rocks do show evidence of water action.

BELOW OLYMPUS MONS
The volcano Olympus Mons is not only the biggest volcano on Mars, but also the biggest in the whole Solar System. The summit caldera (crater) measures more than 50 miles (80 km) across.

Mars is also home to the highest mountains in the Solar System. They are extinct (or perhaps dormant) volcanoes. Three, averaging about 6 miles (10 km) high, are located on a great bulge in the crust called the Tharsis Ridge, which spans the equator. To the northwest lies the magnificent Olympus Mons—15 miles (25 km) high and 400 miles (600 km) across—and the biggest volcano we know in the Solar System.

Just to the east of Tharsis the other outstanding feature of the Martian landscape begins, Valles Marineris—Mars's "Grand Canyon." This great gash in the crust runs just south of the equator for nearly 3,000 miles (5,000 km); in places it is over 250 miles (400 km) wide and 4 miles (6 km) deep. But unlike Arizona's Grand Canyon, which was cut by flowing water, Valles Marineris is a geological fault, formed when the surface fractured long ago.

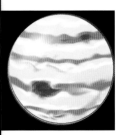

The Giant Planets

With a diameter 11 times that of Earth and with over twice the mass of all the other planets put together, Jupiter is truly gigantic. And Saturn is not that much smaller. Both are made up mainly of gas and liquid gas and are quite unlike the four rocky inner planets.

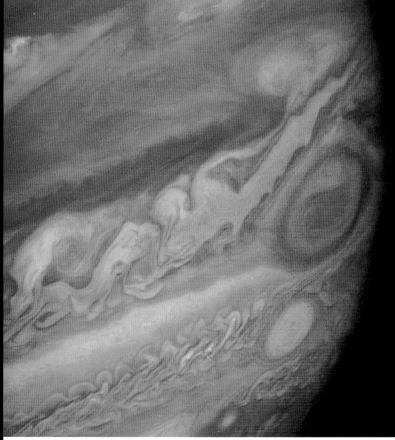

LEFT BELTS AND ZONES
This Voyager image shows clearly the colorful belts and zones that cross Jupiter's atmosphere. The oval regions mark the sites of storms, the most prominent of which is the Great Red Spot.

ABOVE SPOTS AND EDDIES
In this Voyager close-up, gas currents in Jupiter's atmosphere swirl and eddy furiously around the Great Red Spot. This great superhurricane, 17,000 miles (28,000 km) across, has been raging for centuries.

GIGANTIC JUPITER

Jupiter presents a colorful disk, predominantly reddish-orange. It is crisscrossed with alternate dark and pale bands. The dark ones are known as belts, and the pale ones zones. These bands lie parallel to the planet's equator.

What are these bands? They are layers of clouds that have been drawn out parallel by the planet's rapid rotation—Jupiter spins around on its axis in less than 10 hours. No other planet spins so fast.

The bands of clouds course furiously around the planet at different speeds and in different directions. This produces great turbulence at the edges of the bands, creating ever-changing flow patterns, or eddies. Violent hurricane-like storms, pictured as pale or dark ovals, break out all the time. Most persistent is the vividly colored Great Dark Spot.

THE GAS GIANT

Jupiter is unlike Earth and the other terrestrial planets because it has no solid rocky surface. It is made up mainly of hydrogen and helium. These elements are found as gases in Jupiter's deep atmosphere. At the foot of the atmosphere, where the pressure is great, the gases turn into liquid, creating a deep, planet-wide ocean.

Thousands of miles deep in the ocean, pressures build up so high that they compress the hydrogen into a liquid metal-like state, rather like mercury. Only at the very center of the planet is there probably a small rocky core.

LORD OF THE RINGS

Saturn is the second largest planet after Jupiter. And its disk is a pale imitation of Jupiter's, crossed with faint parallel bands of clouds caused by the planet's swift rotation. But the most prominent band on the disk is the dark shadow cast by Saturn's crowning glory, its magnificent ring system.

In composition, Saturn is also similar to Jupiter, made up of hydrogen and helium in gaseous and liquid states. Overall, it is less dense than Jupiter and indeed any other planet, with a relative density of less than 1. This means that Saturn would, if it could, float in water.

RINGS AND RINGLETS

The beautiful rings that surround Saturn's equator span more than twice the planet's diameter, but they are only up to about 0.6 mile (1 km) thick.

From Earth only three rings are readily visible—A (outer), B, and C (inner). Space probes have found several more—a faint D ring inside the C, and F, G, and E rings outside the A.

Close-up pictures show that the rings are made up of thousands of separate ringlets. These mark the paths of the swift-moving icy particles that form the rings.

ABOVE HUBBLE'S SATURN
A glorious picture of Saturn taken in infrared light by the Hubble Space Telescope. False colors have been used to highlight the parallel cloud bands in the atmosphere. The classic A, B, and C rings show up clearly.

ABOVE SATURN'S RING SYSTEM
In all, Saturn's extensive ring system (compared, here, in size with Earth) measures some 170,000 miles (270,000 km) across. The different colors in this false-color image of the rings indicate mainly different sizes of ringlet particles.

Far Distant Worlds

Until March 1781, astronomers knew of only six planets, the farthest out being Saturn. But on the 13th of that month, English musician-turned-astronomer William Herschel discovered another, which came to be called Uranus. Amazingly, it was twice as far away as Saturn.

LEFT URANUS FROM MIRANDA
This is the view you would see of Uranus if you were orbiting its moon Miranda. The planet is overall a bluish-green color, with no features visible in the atmosphere. The color is due to the presence of methane, which selectively absorbs red wavelengths of light, leaving mainly blue ones.

Topsy-Turvey Uranus

If you know exactly where to look, you can just see Uranus (maximum magnitude 5.5) with the naked eye. And you can easily follow it with binoculars. In 2004 it was traveling through Aquarius, and will cross into Pisces in 2009.

The third largest planet after Jupiter and Saturn, Uranus is about four times bigger in diameter than Earth. In telescopes, it presents a bluish green disk, a color confirmed by photographs taken by space-borne telescopes. It has a deep atmosphere of mainly hydrogen and helium above a vast ocean containing water, ammonia, and methane.

Uranus spins around on its axis in about 17 hours, but its axis is tipped right over compared with the other planets. So, in effect, it spins on its side as it orbits the Sun.

In 1977, astronomers discovered that Uranus has a ring system, like Saturn's but very much fainter. It is made up of 11 narrow rings as black as soot.

Deep Blue Neptune

Following Uranus's discovery, astronomers began searching for other planets. And on September 23, 1846, the German astronomer Johann Galle found one, which was called Neptune.

Neptune proved to be a near-twin of Uranus in size and to have much the same make-up. Telescopes show it as a bluish disk. Again, space telescopes confirm this color. But whereas Uranus is uniformly bland in appearance, Neptune shows a variety of features in the atmosphere. They include cloud bands, dark storm centers, and plenty of white cirrus-type clouds.

Like Uranus, Neptune has a faint ring system, consisting of two narrow bright rings, and two broad faint ones. This means that all four gas giants—Jupiter through Neptune—have ring systems.

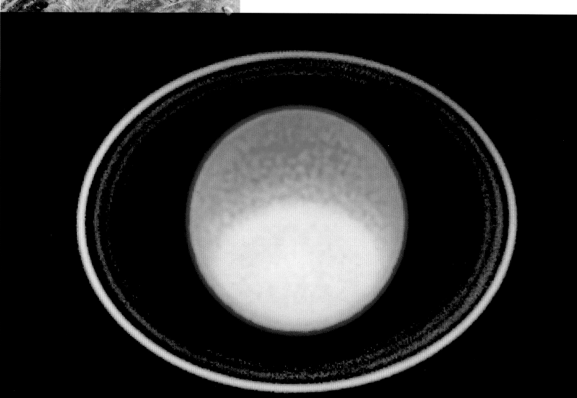

LEFT RINGS AROUND URANUS
Uranus's ring system shows up well in this infrared image taken by the Hubble Space Telescope. Atmospheric haze shows up as the pinkish ring around the edge of the planet. The pale region shows cloud covering most of the south polar region.

FAR DISTANT WORLDS

ABOVE DOUBLE PLANET
Pluto and its moon Charon, pictured by the Hubble Space Telescope. Because Charon is so big, the two bodies are considered to be a double planet. The Hubble Telescope has been able to make out a few vague surface features on Pluto, which spins around once in about six days.

BELOW SCATTERED CLOUDS
When the Hubble Space Telescope began to monitor Neptune in the 1990s, the Great Dark Spot had disappeared. Dark bands were still visible in the atmosphere, and there were flecks of scattered clouds.

A DOUBLE PLANET

After the discovery of Neptune, astronomers became convinced that there were still other planets to find. But it was not until February 18, 1930, that the next one was discovered, by U.S. astronomer Clyde Tombaugh. It was called Pluto.

With a diameter of only 1,429 miles (2,300 km), Pluto is by far the smallest planet—smaller even than our own Moon. It lies so far away that in telescopes it appears like a faint star.

Pluto is a deep-frozen world made up of rock and ice, and its surface is covered with frozen nitrogen and methane. It is quite different from all the other planets and is probably an example of the many ice worlds that exist in the outer Solar System. This has led to its status as a planet being challenged.

Small though it is, Pluto has a moon circling it, called Charon. It is half Pluto's size, which has led to astronomers considering Pluto/Charon to be a double planetary system.

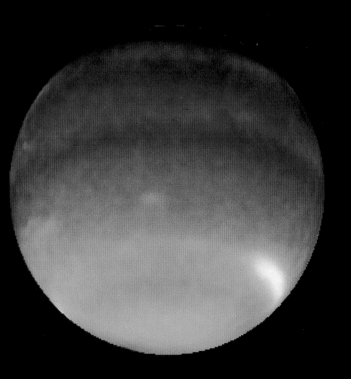
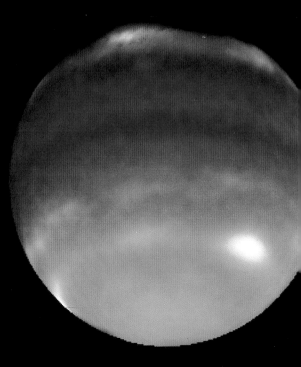

Many Moons

Most planets have one or more natural satellites, or moons, circling around them. They range in size from irregular lumps of rock to bodies bigger than the planet Mercury. The four terrestrial (Earth-like) planets have only three moons between them, compared with more than 130 for the four gas, giant planets.

ABOVE SIX OF THE BEST
The six biggest moons in the Solar System, compared in size of diameter: Ganymede (3,278 miles, 5,275 km) is the largest, followed by Titan (3,200 miles, 5,150 km), Callisto (2,995 miles, 4,820 km), Io (2,257 miles, 3,632 km), the Moon (2,160 miles, 3,476 km), and Europa (1,942 miles, 3,126 km).

MOONS OF EARTH AND MARS

Of the four terrestrial planets, two—Mercury and Venus—have no moons. Earth, of course, has just the one moon—the Moon, which is our closest neighbor in space (see page 122). It is a rocky body with much the same composition as Earth itself. It is exceptionally large for a moon, compared with the size of its parent body.

Mars has two moons, but they are tiny. Phobos is only about 17 miles (28 km) across, while Deimos is even smaller. Shapeless lumps pock-marked with craters, these two rocky bodies are almost certainly former asteroids that were captured by Mars from the nearby asteroid belt.

MOONS OF THE GIANTS

As befits the largest planet, Jupiter has the largest number of known moons—and more are being discovered all the time. By the end of 2004, the number stood at more than 60.

We can see the largest four moons through binoculars. Named the Galilean moons, they are Io, Europa, Ganymede, and Callisto in order of distance from the planet. Ganymede is bigger than the planet Mercury, and Callisto is only a fraction smaller. Io is the most interesting moon because of its active sulfur volcanoes, which account for the moon's garish, pizza-like appearance.

Saturn too has an abundance of moons—more than 30. Its largest, Titan, is also bigger than Mercury and is unique among moons in having a dense atmosphere of nitrogen and methane. Methane clouds float in the atmosphere, and there may even be methane lakes and methane snow on the surface.

Some of the most interesting of Saturn's satellites are the tiny shepherd moons that orbit near the rings. They are so called because they seem somehow to keep the ring particles in place, rather like a shepherd herds sheep.

LEFT IO'S ERUPTIONS
A volcano erupts on the limb of Jupiter's moon Io, blasting out gas and dust to heights of 125 miles (200 km) or more.

Moons of Distant Worlds

The most intriguing of Uranus's 26 moons is Miranda. It has an extraordinary surface, which is a patchwork of quite different geological features butted up together. This might have been the result of a collision with a large asteroid long ago that shattered Miranda to pieces. Then the pieces recombined under gravity to create the weird landscape we see today.

Triton (1,680 miles, 2,700 km) is by far the largest of Neptune's 13 moons. Its surface is covered with frozen methane and nitrogen, which forms a kind of snow near the poles.

The smallest planet, Pluto, has just a single moon, named Charon, discovered in 1978. Some 740 miles (1,190 km) in diameter, it is half the size of Pluto itself. It orbits only about 11,000 miles (18,000 km) from Pluto's surface.

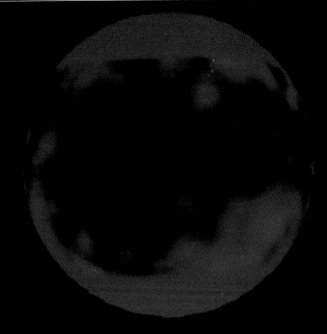

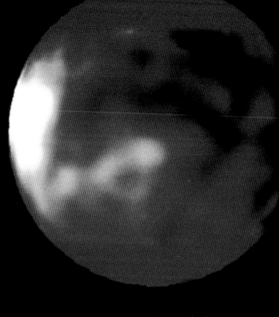

BELOW REFLECTIONS ON ENCELADUS
Saturn's moon Enceladus is one of the most reflective bodies in the Solar System. Some 300 miles (500 km) across, its surface is criss-crossed with faults and peppered with craters.

ABOVE TITAN'S SURFACE
Peering through Titan's dense atmosphere, the Hubble Space Telescope has spotted features on its surface. They could include methane ice cliffs and oceans.

BELOW TRITON'S SNOWSCAPE
Pinkish snow covers the south polar region of Neptune's moon Triton. The dark plumes are subsurface material spewed out by erupting ice volcanoes or geysers.

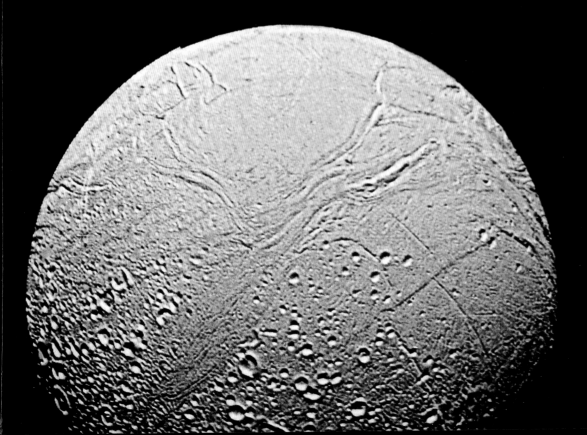

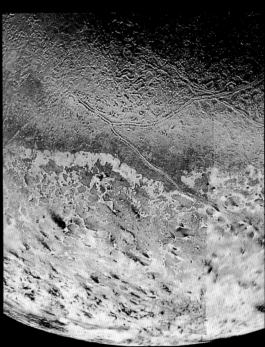

Meteors and Comets

Every clear night, if you remain stargazing for a while, you will almost certainly see what look like stars falling from the sky. They are not falling stars, of course, but meteors—fiery trails made by specks of space dust burning up in the atmosphere.

ABOVE **LEONID METEORS**
Two meteors flash through the sky during the impressive Leonid meteor shower in mid-November 1999. Leonids peak every 33 years, coinciding with the return of their originating comet Tempel-Tuttle.

SHOWERS OF METEORS

Earth is bombarded all the time from outer space by particles we call meteoroids. They are bits of rock and metal, usually little bigger than sand grains.

Attracted by Earth's gravity, meteoroids plunge down into the atmosphere at speeds as high as 45 miles (70 km) a second. Friction with the air particles generates enough heat to make the meteoroids glow white hot and then burn away to dust. They leave behind the fiery trails we see as meteors.

On most nights meteors are sporadic—occurring in any part of the sky at the rate of up to about 10 an hour. But at certain times of the year, the number of meteors rises rapidly, creating what are called meteor showers. And they appear to come from a particular point of the sky, called the radiant.

Meteor showers are named for the constellation in which the radiant occurs. Among the most reliable showers are the Perseids (radiant in Perseus) in August and the Leonids (radiant in Leo) in November.

BELOW **METEOR CRATER**
One of Arizona's natural wonders, Meteor Crater, also called Barringer crater. This great pit in the Arizona Desert measures 4,150 feet (1,265 m) across and is 575 feet (175 m) deep. It was created by the impact of an iron meteorite about 50,000 years ago.

Rocks from Space

Earth attracts something like 220,000 tons (200,000 tonnes) of meteoroids every year. Most of this material burns up in the atmosphere. But some of the larger meteoroids survive their fiery fall and reach the ground—as meteorites.

Their passage through the atmosphere can be spectacular, creating a bright fireball and generating sonic booms. If the meteorites are big enough, they can gouge out huge craters—Meteor Crater in Arizona is a prime example.

Two basic types of meteorites are found, stony and iron. The stony ones are made up of silicates, like many stones on Earth. The iron ones are made up mainly of iron and nickel, sometimes with a little cobalt.

Most meteorites appear to be bits of asteroids (page 154), although some seem to have come from the Moon or Mars.

Comets

Comets are rare visitors to our skies, but are the most spectacular. They can become bright as the brightest stars, remain visible for months, and grow tails millions of miles long. Hale-Bopp was such a comet, hanging in northern skies for months in 1997.

Comets appear huge in the sky, but at their heart is a lump of frozen ice and dust (a "dirty snowball") only a few miles across. There is thought to be a huge reservoir of such icy lumps at the very edge of the Solar System in a region we call the Oort Cloud.

Periodically, one of the lumps leaves the cloud and ventures into the inner Solar System. At first cold and invisible, it starts to become visible only when the Sun's heat begins to evaporate some of the ice. A cloud of gas collects around the icy lump, along with dust which has been released.

This cloud reflects sunlight, and so the comet becomes visible. The "pressure" of the solar wind "blows" the cloud away from the head of the comet to create a tail, or more usually two—a gas (or ion) tail and a dust tail.

A "new" comet like this can take thousands of years to travel around the Sun and back. Other comets travel around the Sun in much shorter periods, and reappear regularly in Earth skies. We call them periodic (P/) comets. Best known is P/Halley's Comet, which returns about every 75 years. Last seen in 1986, it should return in 2061.

ABOVE **Halley's Nucleus**
Bright jets of gas spurt from the nucleus of Halley's comet in March 1986. This picture was returned by the European space probe Giotto. The nucleus, the darker region at top left, appears to be about 9 miles (15 km) long and about half this size across.

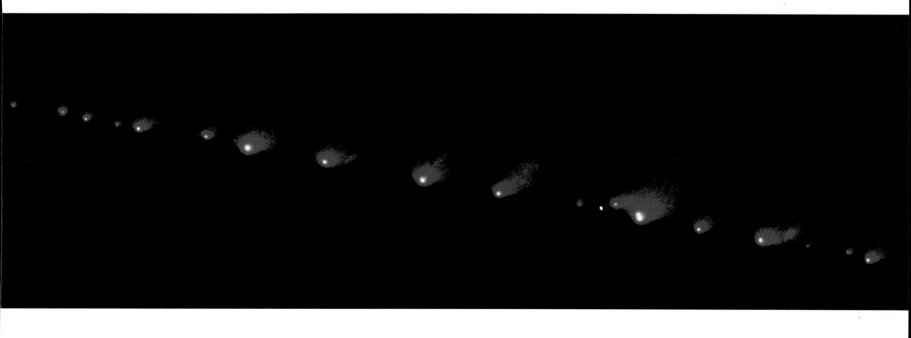

BELOW **Collision Course**
The string of fragments remaining after Comet Shoemaker-Levy 9 had been ripped apart by Jupiter's gravity in July 1993. Here we see them in May the following year, heading on a collision course with Jupiter. The fragments began bombarding the planet on July 16, 1994, creating huge fireballs in the atmosphere.

The Asteroids

Billions of space rocks, large and small, circle the Sun between the orbits of Mars and Jupiter. They are the asteroids, also named the minor planets. They are probably the leftovers from the early days of planet formation. Another planet couldn't form in this region because of the powerful gravitational disturbances from Jupiter.

The Asteroid Belt

Italian astronomer Giuseppe Piazzi discovered the first asteroid, Ceres, on the first night of 1801. Over the next six years, three more were discovered—Pallas, Juno, and Vesta. Since that time tens of thousands have been discovered and had their orbits computed.

Most asteroids orbit within a broad band known as the asteroid belt, which spans a region between about 150–370 million miles (250–600 million km) from the Sun. Within this band, the asteroids take between three and six years to circle the Sun.

What Asteroids are Like

The asteroids are termed "minor" planets, because they are very small indeed compared with the true planets. Even the largest, Ceres, measures only about 580 miles (933 km) across. And there are only about 10 larger than 150 miles (250 km) across.

Only Ceres, Pallas, Vesta and a few other asteroids are roughly spherical in shape. The others are irregular-shaped lumps, looking rather like huge potatoes. From the way they reflect light, asteroids appear to be of three main kinds. Some are made up mainly of rock, some mainly of metal, and others a mixture of the two.

No one knew what asteroids looked like until the space probe Galileo photographed Gaspra in 1991. It proved to be pock-marked with craters, like the other asteroids that have since been photographed, such as Eros.

Observing Asteroids

Because they are so small and so far away, the asteroids—with one exception—can't be seen with the naked eye. The exception is Vesta (diameter about 340 miles, 550 km), which shines at a maximum magnitude of about 5.2. At magnitude 6.7, Ceres is just outside naked-eye range.

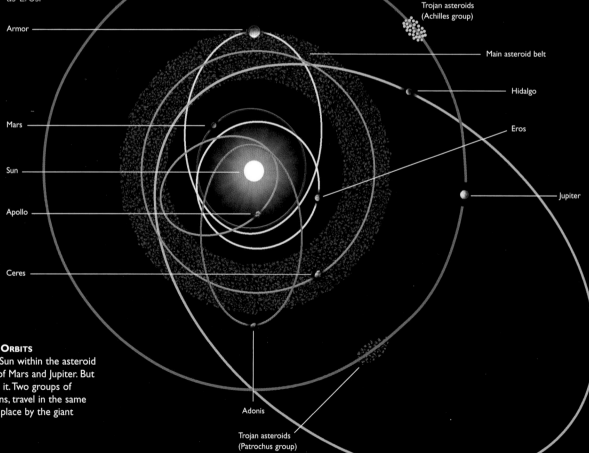

ABOVE RIGHT ASTEROID ORBITS
Most asteroids orbit the Sun within the asteroid belt, between the orbits of Mars and Jupiter. But some stray far away from it. Two groups of asteroids, called the Trojans, travel in the same orbit as Jupiter, locked in place by the giant planet's gravity.

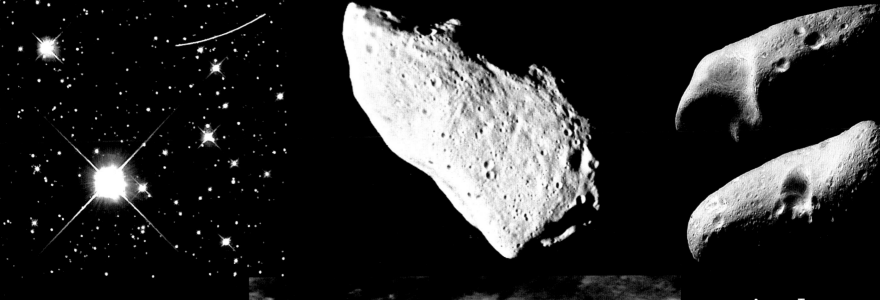

The path of Vesta and other bright asteroids can readily be followed through binoculars. You can find out their location at any time from tables in astronomical magazines and yearbooks.

If you have a driven telescope, you can easily detect asteroids in long-exposure photographs. They show up as trails against the background of stars because they are moving comparatively quickly.

NEAR-EARTH OBJECTS

As the diagram (left) of asteroid orbits shows, many asteroids wander in and out of the asteroid belt—some head closer in toward the Sun, others out beyond Jupiter's orbit.

Of particular interest to us humans are the errant asteroids we call near-Earth objects (NEOs). These are asteroids whose orbits take them uncomfortably close to Earth and pose the potential threat of collision with our planet.

We know from fossil records that asteroid collisions have occurred in the past. One that happened 65 million years ago was responsible for the mass extinction of living species, including the dinosaurs. Early in 2002, NEO 2002 NY40—visible through powerful binoculars—passed within 300,000 miles (500,000 km) of Earth, showing that the threat from space is frighteningly real.

ABOVE LEFT ASTEROID TRAIL
A curving asteroid trail shows up in a long-exposure photograph of the stars in the southern constellation Centaurus taken by the Hubble Space Telescope. Because asteroids can interfere with astronomical observations, they are sometimes called the "vermin of the skies."

ABOVE CENTER ASTEROID GASPRA
The space probe Galileo took this picture of asteroid Gaspra in October 1991. About 12 miles (20 km) long, this rocky asteroid circles around the Sun every 3.3 years.

ABOVE RIGHT ASTEROID EROS
The two hemispheres of asteroid Eros, pictured by the probe NEAR-Shoemaker in 1999. The next year, on Valentine's Day (an appropriate date for a probe named after the son of the Greek goddess of love), Eros went into orbit around the asteroid and circled it for a year before landing on it on February 12, 2001.

LEFT PSYCHE CRATER
From its orbit 30 miles (50 km) high, NEAR-Shoemaker took this close-up picture of the floor of a large crater on Eros named Psyche. The surface has a thin coating of soil, peppered with tiny craters.

Glossary

Absolute magnitude
A measure of the true brightness of a star. It is what the brightness of the star would be if viewed from a distance of 10 parsecs, or 32.6 light-years.

Active galaxy
One that has an exceptionally high energy output.

Antoniadi scale
A scale used by an observer to indicate the quality of seeing.
I   Perfect seeing
II Mainly calm, but slight disturbance
III Moderate seeing, more disturbance
IV Poor seeing, constant disturbance
V   Very bad seeing, very bad disturbance

Apparent magnitude
A measure of the brightness of a star as we see it from Earth.

Asteroids
Small rocky bodies that circle the Sun, mainly in a "belt" between the orbits of Mars and Jupiter. Also called the minor planets and planetoids.

Astrology
Study of the positions of the heavenly bodies with a view to explaining events in people's lives or foretelling the future.

Astronomical unit (AU)
The distance between the Earth and the Sun, 93,000,000 miles (149,600,000 kilometers).

Astronomy
The scientific study of the universe and the bodies therein – the Sun, the Moon, the planets, their moons, the stars, and the galaxies.

Aurora
A colorful glow seen mainly in far northern and far southern skies, produced when charged particles from the Sun collide with particles from the upper atmosphere. In northern skies it is called aurora borealis, or Northern Lights; and in southern skies the aurora australis, or Southern Lights.

Big Bang
The name for the event that is believed to have created the universe, some 15,000 million years ago.

Binary
A two-star system in which the stars are associated and circle around a common center of gravity (barycenter).

Black hole
A region of space in which gravity is so immensely strong that not even light can escape from it.

Celestial sphere
An imaginary sphere around the Earth, to the inside of which the stars seem to be fixed.

Cepheids
A common class of variable stars that vary in brightness as regularly as clockwork. They care named after the prototype star Delta Cephei.

Comet
A small member of the solar system, a "dirty snowball" of rock, ice, and dust, which starts to shine when it approaches the Sun.

Conjunction
The lining up of heavenly bodies in the sky, such as a planet, the Sun, and the Earth.

Constellations
Imaginary patterns that the bright stars make in the night sky.

Cosmology
The study of the origin, evolution, and structure of the universe.

Culmination
The maximum altitude of a heavenly body above the horizon; this happens when it crosses the meridian.

Declination
A star's celestial latitude on the celestial sphere, measured in degrees north (+) or south (−) of the celestial equator.

Double star
A pair of stars that appear close together in the sky. In a binary star, the two components are physically associated. In an optical double, the components are usually quite separate, appearing together only because they lie in the same direction.

Eclipse
The passing of one heavenly body in front of another, blotting out its light. A solar eclipse, or an eclipse of the Sun, occurs on Earth when the Moon passes in front of the Sun. A lunar eclipse, or an eclipse of the Moon, occurs when the Moon moves into the Earth's shadow in space.

Eclipsing binary
A binary-star system that varies in brightness when its two components periodically pass in front of one another in our line of sight.

Ecliptic
The apparent path of the Sun through the heavens during the year.

Equinoxes
Times of the year when the lengths of the day and the night are equal all over the world because the Sun is directly over the Equator. This happens twice: on March 21 (the vernal, or spring equinox); and on September 23 (the fall equinox).

Evening star
A planet (usually Venus) that shines brightly in the western sky at sunset.

Expanding universe
The theory that the universe is getting bigger. Astronomers believe that expansion began with the Big Bang.

Fireball
An exceptionally bright meteor; also called a bolide.

Galaxy
A star "island" in space. Galaxies are usually elliptical or spiral in shape. We usually call our own galaxy, the Galaxy, or the Milky Way.

Globular cluster
A globe-shaped group of stars numbering up to a million or more.

Gravity
The force with which the Earth attracts any object near it. The other heavenly bodies exert a similar force. Gravity is the force that literally holds the universe together.

Librations
Oscillations of the Moon as we see it from Earth that enable us to see slightly "around the edges" of the face it presents.

Light-year
A common unit for measuring distances in space. It is the distance light travels in a year: 5.88 million million miles (9.46 million million kilometers).

Limb
The edge of the visible disk of a planet or a moon.

Luminosity
A measure of the total energy given off by a star.

Lunar
Relating to the Moon.

Magnitude
The scale on which star brightens is measured. On the scale, the visible stars are divided into six levels of brightness, and the scale is extended to describe brighter and dimmer stars. See also *Absolute magnitude*; *Apparent magnitude*.

Mare
A large plain on the Moon. "Mare" (plural "maria") is the Lain word for "sea". The dark areas we see on the moon are maria. The main sea areas (and English names) are:
Lacus Mortis (Lake of Death)
Lacus Somniorum (Lake of Dreams)
Mare Crisium (Sea of Crises)
Mare Fecunditatis (Sea of Fertility)
Mare Frigoris (Sea of Cold)
Mare Humorum (Sea of Humors)
Mare Imbrium (Sea of Showers)
Mare Nectaris (Sea of Nectar)
Mare Nubium (Sea of Clouds)
Mare Orientale (Eastern Sea)
Mare Serenitatis (Sea of Serenity)
Mare Smythii (Smyth's Sea)
Mare Tranquillitatis
 (Sea of Tranquility)
Mare Vaporum (Sea of Vapors)
Oceanus Procellarum
 (Ocean of Storms)
Sinus Iridum (Bay of Rainbows)
Sinus Medii (Central Bay)
Sinus Roris (Bay of Dews)

Meridian
The great circle on the celestial sphere that passes through the north and south poles. An observer's meridian passes through the north and south points on the horizon and the zenith.

Meteor
A streak of light we see when a piece of rock from outer space plunges through the atmosphere and burns up.

Meteorite
A piece of rock from outer space big enough to survive its passage through the atmosphere and reach the ground.

GLOSSARY

Milky Way
A pale band of light that we can often see arching across the heavens. It is a region dense with stars and is, in effect, a cross-section of our Galaxy, which is also called the Milky Way.

Moon
A natural satellite of a planet

Morning star
A planet (usually Venus) that shines brightly in the eastern sky at dawn.

Nadir
The point on the celestial sphere directly beneath an observer.

Nebula
A cloud of gas and dust between the stars. We see bright nebulae because they glow or reflect light; we see dark nebulae because they obscure light from distant stars.

Neutron star
An incredibly dense body made up of a solid mass of neutrons. See also *Pulsar*.

Nova
A "new" star; actually one that suddenly increases greatly in brightness.

Nuclear fusion
The joining together of the nuclei (centers) of light atoms, such as hydrogen, to make heavier ones, such as helium. It is the energy-production process in stars.

Occultation
The temporary hiding of one heavenly body by another, for example of a star or a planet by the Moon.

Open cluster
A loose grouping of stars, typically containing a few hundred members.

Opposition
The position of a planet in its orbit when it lies exactly opposite the Sun in the sky.

Parallax principle
The apparent change in position of a nearby object against a distant background when looked at from different angles. The parallax of a star is the angle subtended at the star by the Earth's radius.

Parsec
The unit professionals generally use to measure distances in space. It is the distance at which the parallax of a star would be one second of arc (1/3600th of a degree). Equal to 3.26 light-years.

Phases
The different shapes of the Moon we see in the sky during the month, as more or less of its surface is lit up by the Sun. Venus shows noticeable phases as well.

Planets
Large bodies that circle in space around the Sun. Some of the other stars probably have planets circling around them.

Precession
A slow change in the direction in which the Earth's axis points in space, which over the years alters the positions of the stars on the celestial sphere.

Proper motion
Observed motion of a nearby star against the stellar background.

Pulsar
A rapidly rotating neutron star that gives off pulses of radiation.

Quasar
A remote body that looks like a star but has the energy output of many galaxies. Also called quasi-stellar object (QSO).

Radial motion
Motion of a star toward or away from us, detectable from its spectrum.

Retrograde motion
Motion of a heavenly body in the opposite direction from usual.

Right ascension
A star's celestial longitude, expressed usually in hours and minutes of sidereal time. It is measured from the point (First Point in Aries) where the ecliptic meets the celestial equator at the vernal equinox.

Saros
A period of a little over 18 years (6,585.3 days), after which the Moon returns to exactly the same position in relation to the Sun and the Earth. It therefore marks the interval between successive eclipses of the same type.

Satellite
A small body that circles around a larger one in space; a moon. Most of the planets have natural satellites. The Earth has one, the Moon. Saturn has more than 20. Earth also now has thousands of artificial satellites circling around it, launched into orbit by space scientists.

Seasons
Periods of the year marked by noticeable changes in the weather, especially in temperature. They occur because of the tilt in the Earth's axis with respect to the plane of its orbit around the Sun.

Seeing
The quality of the observing conditions at the time of observation.

Sidereal time
Time measured in relation to the stars, based on the sidereal day, the Earth's true period of rotation on its axis, 23 hours 56 minutes 4 seconds of ordinary clock time.

Solar system
The family of the Sun that travels through space as a unit. At the center is the Sun, around which circle nine planets (including the Earth), the asteroids, and numerous comets.

Solstices
Times of the year when the Sun reaches its highest and lowest points in the sky at noon. In the Northern Hemisphere, it reaches its highest point on about June 21 (summer solstice), and its lowest on December 21 (winter solstice). The dates are reversed in the Southern Hemisphere.

Spectrum
A band of color (spread of light wavelengths) produced when light is passed through a spectroscope. Stellar spectra are crossed by dark lines.

Speed of light
Light travels in a vacuum at a speed of about 186,000 miles (300,000 kilometers) a second. It is the fastest speed possible.

Star
A gaseous body that produces energy by nuclear fusion. It releases this energy as light, heat, and other radiation. The Sun is our local star.

Sunspot
An area on the Sun's surface that is darker and cooler than normal. Sunspots come and go regularly over a period of about 11 years.

Supernova
A gigantic stellar explosion in which a supergiant star blasts itself apart.

Supernova remnant
An expanding shell of gas resulting from a supernova explosion.

Terminator
The boundary between the sunlit and dark halves of the Moon or a planet.

Transit
The movement of one small heavenly body in front of a larger one, such as Venus across the disk of the Sun.

Universe
Everything that exists: the Earth, the Moon, the Sun, the planets, the stars, and even space itself.

Variable star
One that changes in brightness. Intrinsic variables (such as Cepheids) vary in brightness because of internal physical changes. Extrinsic variables (such as eclipsing binaries) vary because of some external influence.

Zenith
The point on the celestial sphere directly above an observer.

Zodiac
An imaginary band in the heavens in which the Sun, Moon, and planets are always found. It is occupied by 12 constellations, the constellations of the zodiac.

Index

Constellations are shown in capitals. Page numbers in bold refer to main entries. Page numbers in italics refer to maps.

Deep-sky objects appearing only on maps but not in text or captions are omitted from the index, as are lunar features appearing only in the quadrants.

3C-273 42, 90
30 Doradus see Tarantula Nebula
47 Tucanae 35, 72, 74, 98

A

Abell 2256 18
Achernar 62, *62*, 112
Al Nath (Beta Tauri) 106, 108, 114
Albategnius plain 134
Albireo 102
Alcor 34, 84
Alcyone *34*, 114
Aldebaran 35, 51, 56, *56*, 62, *62*, 76, 110, 114
Algeiba 82
Algol 34, 89, 106, 113
Aliacensis crater 134
Alioth 84
Alkaid 84
Alphard 82
Alpheratz (Alpha Andromedae) 104, 106, 108
Alphonsus plain 130
Alpine Valley 124, 132
Alps, lunar *124*, 128
Altair 58, 60, 92, 96, 100, 101
ANDROMEDA 58, **108**, 111
   51 Andromedae 106
   Alpha see Alpheratz
   Beta see Al Nath
   Epsilon see Enif
Andromeda Galaxy *38*, 39, 40, 44, 45, 108, 111
Ant Nebula 32, 33
Antares 58, *58*, 60, 92, 94, 97
Aphrodite Terra 143
Apollo explorations *123*, *127*, 130, *132*, *132*
Appenines 128, 132, *133*
April stars 86–87
AQUARIUS **104**, 106
AQUILA 36, **100**, 101, 102
   Alpha see Altair
ARA **72**
Arches 35
Archimedes crater 128
Arcturus 50, 58, *58*, 60, *60*, 88, 92
Arecibo radio telescope 18
Ares Vallis *145*

Argyre basin 144
ARIES **110**
Aristarchus crater 128, 129
Aristillus crater 132
Aristoteles crater 132
asteroids 137, 150, 152, **154–155**, *154*, 155
astronomical unit 118
Atlas 114
Atlas crater 132
August stars 100–103
AURIGA 36, **76**
   see also Capella
aurora australis 118
aurora borealis 118, *118*
Autolycus crater 132

B

Bailly crater 130
Barnard 33 see Horsehead Nebula
Bayer system 66
Beehive cluster see Praesepe cluster
Bessel crater 132
Betelgeuse *51*, 56, 76, 78, 114
BIG DIPPER *51*, 50, 56, 68, 84, 86, 88
   see also URSA MAJOR
binary stars 25, 34
binoculars 13
BL Lacertae 42, 104
Black-Eye Galaxy 88
black holes 33
   supermassive 37, 43, 90, *90*
   blazars 42, 104
BOÖTES **88**
   Alpha see Arcturus
brown dwarfs 24
Bubble Nebula 70
Bulliadus crater 130
Butterfly Cluster 94

C

Callisto 150, *150*
Caloris Basin 142
CANCER **80**
CANES VENATICI **86**
   Alpha see Cor Caroli
CANIS MAJOR **76**
   Alpha see Sirius
CANIS MINOR 56, **80**
   Alpha see Procyon
Canopus 24, 62, 72, 74
Capella 50, 51, 62, 76, 77, 112
CAPRICORNUS **100**
captured rotation 124
CARINA 36, **72**, 83
   Eta 72, *72*
   stars within system 51, 60, 72;
   see also Canopus
Carpathians, lunar 128

CASSIOPEIA 36, 56, 68, **70**, 104
   stars within system 68, 70
Cassiopeia A 33
Castor 50, 51, 62, 76, 112
Catharina crater 134
Cat's Eye Nebula 33, 68, *68*
Caucasus Mountains, lunar 128, 132, *133*
celestial sphere 53–53, *52*, 54, 66
   latitude and longitude 53, 66
   poles 65
CENTAURUS 36, 50, **74**, 88, 92
   Alpha 51, *51*, 60, *60*, 62, 74, 112
   Beta 51, *51*, 60, *60*, 62, 74
   Omega 35, 74, *74*, 98
   Proxima 25, 74
   R variable 74
Centaurus A 42, *42*
Cepheid variables 25, 68, 100
CEPHEUS **68**
   Mu see Garnet Star
Ceres 154
CETUS 62, 106, **110**
   Omicron see Mira
Charon 149, *149*, 151
Circlet 106
Clavius crater 130, *131*
Cleomenes crater 132
clusters 45
   globular 35, 37, 41, 66
   open 35, 66
   see also individual clusters, e.g. under M numbers or NGC numbers
Coal Sack 29, 74
COLUMBA **76**
COMA BERENICES **88**, 90
   Beta 86
Coma cluster 45, 88
Coma-Virgo cluster 90
cometary knots *105*
comets 13, 17, 36, 137, **153**, *153*
Cone Nebula 80
constellations 48, *49*
   circumpolar 65, 68—75, 84
   Latin–English list 49
   seasonal views 56–63
   of zodiac see individual constellations, e.g. CANCER
Copernicus, Nicolaus 139
Copernicus crater *124*, 126, 128, *129*
Cor Caroli 86
Cor Hydrae see Alphard
Cor Leonis see Regulus
CORONA BOREALIS 60, **92**
CORVUS **86**, 90
Crab Nebula 114
Crimson Star 77
CRUX 36, 51, *51*, 60, *60*, 62, 72, **74**

CYGNUS 36, 48, 56, 100, **102**, 104
   Alpha see Deneb
   Beta see Albireo
   61 Cygni 102
Cygnus Loop 102
Cygnus Rift 29, 102
Cyrillus crater 134

D

December stars 112–115
declination 53, 66
deep-sky objects 66;
   see also individual objects, e.g. under M numbers or NGC numbers
Deimos 150
DELPHINUS **100**
Deneb 25, 50, 58, 60, 92, 96, 100, 102
Denebola 82
DORADO 38, 62, **72**
Double Cluster 113
double-double star 96
double planetary systems 122, 149
double stars see binary stars
DRACO **68**
   Alpha see Thuban
Dubhe 50, 68, 84
Dumbbell Nebula 01

E

Eagle Nebula 28, *30*, 97
Earth, movement of 54, *54*
eclipses: lunar 120, 124, *124*
   to photograph 17
   solar 118, 120–121, *120*, *121*, 124
   total 120, *120*, 121, *121*
eclipsing binaries 25, 34
ecliptic 52, *53*, 54
Electra 114
electromagnetic waves 18–19
Enceladus *151*
Enif 106
Eratosthenes crater 128
ERIDANUS 62, 100, **112**
   Alpha see Achernar
Eros 154, *155*
Eskimo Nebula 33, 80, *81*
ESO 510-G13 *41*
Eta Carinae Nebula 65, 72, *72*
Eudoxus crater 132
Europa 150, *150*
European Southern Observatory 14–15
extrasolar planet 106

F

False Cross 51, 60, 72, 83
February stars 80–81
Fomalhaut 62, 105, 107
FORNAX 107, **112**

Fornax A 112
Fornax Cluster 112
Fornax System 112
Fra Mauro formation 130
Fracastorius crater 134

G

galaxies **40–45**, 66
   active 42–43
   barred-spiral 40, *40*
   dwarf 112
   elliptical 40, *40*, 41, *43*
   interaction of 44–45, 87, 114
   irregular 38, 40
   Messier 90
   satellite 38, 39
   spiral 36, 39, 40, *40*, 41
   starburst *41*, 44
   see also individual galaxies, e.g. Andromeda Galaxy
Galileo Galilei 11
Galle, Johann 148
Ganymede 150, *150*
Garnet Star 68
Gaspra 154, *155*
Gassendi plain 130
Gauss crater 132
GEMINI 56, 76, **80**
   see also Castor; Pollux
Gemini North Telescope 14
Ghost of Jupiter Nebula 87
giant molecular clouds 30
Great Nebula see Orion Nebula
Great Red Spot *146*, 147
Great Spiral see Andromeda Galaxy
Grimaldi crater 126, 130
GRUS **104**

H

Haemus Mountains 132
Hale-Bopp Comet 153
Hale Telescope 14
Halley, Edmond 86
Halley's Comet 36, 153, *153*
Heidi 76
Helix Nebula 104, *105*
Hellas basin 144
HERCULES **92**, 96
   stars within system 35, 86, 92
Hercules crater 132
Herschel, William 68, 148
Hipparchus 24
Hipparchus plain 134
Hodge 301, *11*
Hooker Telescope 40
Horsehead Nebula 29, *29*, 78
H-R (Hertzsprung-Russell) Diagram 26, 27
Hubble, Edwin 20, 40

Hubble Space Telescope 19, **20–21**, *20*, 40
Hyades 35, 110, **114**
HYDRA (Head) **82**
 (Tail) **86–87**
Hyginus Rille 132

I
IC 434 see Horsehead Nebula
IC 2163 45
IC 2391 83
IC 2602 72
Io 150, *150*
Ishtar Terra 143

J
Jacobus Kapteyn Reflector *14*
James Webb Space Telescope 21
January stars 76–79
Jewel Box cluster 74, *74*
Job's Coffin 100
July stars 96–99
June stars 92–95
Jupiter 137, 139, 140, *140*, 141, **146–147**, *146*
 moons 150, *150*, 151
Jura Mountains, lunar 128

K
Keck reflectors 14
Kepler, Johann 139
Kepler crater 128, *129*
Kitt Peak National Observatory *118*

V
Lacaille 9352 105
LACERTA **104**
Lacus Mortis 132
Lacus Somniorum 132
Lagoon Nebula 28, 31, 98
Langrenus crater 134, *134*
Lansberg crater 128
Large Magellanic Cloud *11*, 36, **38**, 44, 62, 72
LEO 56, 58, **82**, 86
 Alpha see Regulus
 Beta see Denebola
 Gamma see Algeiba
LEPUS **77**
 R Leporis see Crimson Star
LIBRA **89**, 94
 Alpha see Zubenelgenubi
 Beta see Zubenelchemale
 Delta 89
light-year 24
Local Group 39, 45, 108, 111
Local supercluster 45
long-period variables 81, 84, 110
 see also Mira variables

Longomontanus crater 130
LUPUS **92**
LYNX **82**
LYRA **96**
 Alpha see Vega

M
M1 see Crab Nebula
M2 104
M3 86, *87*, 88
M4 94, *94*
M5 93
M6 see Butterfly Cluster
M7 94
M8 see Lagoon Nebula
M10 97
M11 see Wild Duck cluster
M12 97
M13 35, 86, 92
M15 106, *107*
M16 97
M17 see Omega Nebula
M18 98
M19 97
M20 see Trifid Nebula
M22 98
M23 98
M24 see Small Sagittarius Star Cloud
M27 101
M29 102
M30 100
M31 see Andromeda Galaxy
M32 38, 39, 108, *108*
M33 see Triangulum Galaxy
M35 80
M36 76
M37 76
M38 76
M39 102
M41 76
M42 see Orion Nebula
M43 78
M44 see Praesepe
M45 see Pleiades
M48 82
M51 see Whirlpool Galaxy
M52 70
M54 98
M58 90
M59 90
M60 90
M63 86
M64 see Black-Eye Galaxy
M65 82
M66 82
M67 80
M69 98
M70 98
M71 101

M72 104
M74 19, 107
M77 110
M80 35
M81 84
M82 84
M83 87
M87 21, 42, *43*, 90, *90*
M89 90
M90 90
M92 92
M94 86
M95 82
M96 82
M97 see Owl Nebula
M100 88, *89*
M101 84
M103 70
M104 see Sombrero Galaxy
M106 86
M108 84
M110 108
Magellanic Clouds 39, 45, 72
 see also Large Magellanic Cloud; Small Magellanic Cloud
Maginus crater 130
magnitude 24–25, 27
Maia 114
Main Sequence stars 26, 30
Manilius crater 132
March stars 82–85
Mare Australe 134
Mare Cognitum 130, *131*
Mare Crisium 132, *133*
Mare Fecunditatis 126, 132, 134, *135*
Mare Frigoris *129*, 132
Mare Humorum *129*, 130, *131*
Mare Imbrium *124*, 126, 128, *129*, 132, *132*
Mare Nectaris 126, 134, *135*
Mare Nubium 130, *131*
Mare Serenitatis 126, 128, 132, *133*
Mare Smythii 134
Mare Tranquillitatis 126, *127*, 132, *132, 133*
Mare Vaporum 132
Mars 94, 137, 139, 140, *140*, **144–145**, *144, 145*
 moons 150
Maskelyne Rille *132*
Mauna Kea Observatory 14
Maurolycus crater 134
May stars 88–91
Megrez 84
Menelaus crater 132
Merak 50, 68, 84
Mercury 137, 139, 140, **142**, *142*, 150
meridian 52, 65, 88, 100, 106

Merope 114
Messier, Charles 66, 90, 98, 108, 114
Messier numbers 66–67
Meteor Crater, Arizona *152*, 153
meteors/meteorites 12, 137, **152–153**, *152*
Milky Way 13, 23, **36–37**, *36, 37, 83*, 96, 100, 104, 114
Mira 110
Mira variables 25, 70, 74, 77, 78, 86, 102, 108
Miranda *148*, 151
Mirfak 112
Mizar 34, 84
Moon 12, 13, 117, 118, 120–121, **122–127**, 150, *150*
 craters 122, *123*, 126, *126*
 distance 122
 maria 122, *123*, 126
 Northeast Quadrant 132–133
 Northwest Quarant 128–129
 phases 122, 124, *124*
 to photograph 16, *16*
 rilles and ridges 126
 rocks *127*, 134
 size 122
 Southeast Quadrant 134–135
 Southwest Quadrant 130–131
moons of other planets *148*, 149, *149*, **150–151**, *150, 151*
Mount Palomar Observatory 14
Mount Wilson Observatory 40
multiple stars 34, 80, 94
Mz3 see Ant Nebula

N
nadir 52
near-earth objects 155
nebulae **28–29**, 66
 bright 28
 dark 29, 30, 74
 planetary 32, 33, 68, 84, 87, 96, 101, 104, 108
 see also individual nebulae, e.g. Ant Nebula
Needle Galaxy 88
Neptune **148**, *149*
 moons 151, *151*
neutron stars 23, 33
New General Catalog numbers 67
Newton crater 126
NGC 205 38, 39, 108
NGC 253 106
NGC 404 108
NGC 457 70
NGC 604 111
NGC 654 70
NGC 663 70
NGC 752 108

NGC 869 see Double Cluster
NGC 884 see Double Cluster
NGC 891 108
NGC 1316 see Fornax A
NGC 1365 112
NGC 1512 113
NGC 1851 76
NGC 2070 see Tarantula Nebula
NGC 2207 45
NGC 2244 see Rosette Nebula
NGC 2264 see Cone Nebula
NGC 2362 76
NGC 2392 see Eskimo Nebula
NGC 2451 81
NGC 2477 81
NGC 2516 72
NGC 2591 83
NGC 3242
 see Ghost of Jupiter Nebula
NGC 3310 41
NGC 3766 74
NGC 4314 21
NGC 4414 37
NGC 4565 see Needle Galaxy
NGC 4751 101
NGC 5128 see Centaurus A
NGC 5194 87
NGC 5195 87
NGC 5286 74
NGC 5466 88
NGC 5822 92
NGC 5986 92
NGC 6210 see Turtle Nebula
NGC 6231 94
NGC 6397 72
NGC 6543 see Cat's Eye Nebula
NGC 6992 see Veil Nebula
NGC 7000
 see North America Nebula
NGC 7009 see Saturn Nebula
NGC 7243 104
NGC 7293 see Helix Nebula
NGC 7635 see Bubble Nebula
North America Nebula 28, 29, 102
NORTHERN CROWN
 see CORONA BOREALIS
northern molecular belt 78
novae 25
November stars 110–111

O
Oceanus Procellarum 126, 128, *128*, *129*, 130, *130*, 132
October stars 106–111
Olympus Mons 145, *145*
Omega Nebula 98
Oort Cloud 153
OPHIUCHUS 93, **96–97**
optical doubles 34, 68, 74

ORION 48, *49*, *50*, 56, 62, 76, **78**, 82, 110, 112
  stars within system 48, 50–51, *51*, 78;
  see also Betelgeuse; Rigel
Orion Nebula 24, 28, *28*, *51*, 78, *78*
Otto Struve crater 128
Owl Nebula 84

P

Parkes radio telescope 18
Paranal Observatory 14–15
parsec. 24
PEGASUS 56, 58, **106**, 110
  Alpha 104
  Square of 56, 104, 106, 108, 110
  stars within system 104, 106
PERSEUS 36, **112–113**
  Alpha see Mirfak
  Beta see Algol
Petavius crater 134, 135
Phad 84
Phobos 150
PHOENIX **106**
photography 15, 16–17
Piazzi, Giuseppe 154
Piccolomini crater 134
Pico mountain 128
PISCES **106–107**, 110
PISCIS AUSTRINIS 104, **105**, 107
Pitatus crater 130
planets 137
  movement of *138*, 140
  see also individual planets
planispheres *13*
Plato crater *124*, 126, 128
Pleiades 34, 35, 51, *56*, 62, 70, 110, **114**
Pleione 114
Plinius crater 132
Plow see Big Dipper
Pluto 137, 139, *139*
  moons 149, *149*, 151
Polaris (Pole Star) *16*, 50, *54*, 68, 70, 84
Pollux 50, 51, 56, 62, 76, 112
Posidonius crater 132
Praesepe cluster 80
Procyon 51, 56, 62, 76, 80, 112
protostars 30
Ptolemaeus crater 130, *131*
Ptolemaic system 138–139
pulsars 18
PUPPIS 36, **81**

Q

quasars 18, 42, *42*, 90

R

radio astronomy 18
radio galaxies 42, *42*
red dwarf 112
red giants 25, 32, 68, 88, 92, 106, 108, 110, 114
Regulus 50
Reinhold crater 128
Rheita crater 134
Riccioli crater 130
Rigel 24, 25, *51*, 56, 76, 78, 112
right ascension 53, 66
Rigil Kent see CENTAURUS, Alpha
Ring Nebula 33, 96, 97
Roque de los Muchachos Observatory 14, *15*
Rosette Nebula 28, 80

S

Sabaru Telescope 14
SAGITTA **101**
SAGITTARIUS 36, 58, 60, **98**, 100
Sagittarius A 37
Sagittarius dwarf galaxy 38, 44
Sagittarius star cloud 23
Saturn 137, 137, 139, 140, *140*, 141, *141*, 146, **147**, *147*
  moons 150, *150*, *151*
Saturn Nebula 104
Schickard crater 130
SCORPIUS 34, 48, 58, 60, 89, 92, **94**, 96
  Alpha see Antares
SCULPTOR **107**
SCUTUM **97**
Sedna 139
September stars 104–145
SERPENS CAPUT **92**, 96, 97
SERPENS CAUDA 92, 96, **97**

Seven Sisters see Pleiades
Seyfert galaxies 42, *43*, 110
Seyfert's Sextet *44*
Shoemaker-Levy Comet *153*
Sickle 82
sidereal time 53, 66
Sinus Iridum *124*, 128
Sirius 24, 25, 51, 56, 62, 66, 74, 76, 112
Sirius B 76
Small Magellanic Cloud 36, **38**, 44, 62, 72
Small Sagittarius Star Cloud 98
Solar System 122, 137, **138–139**, 153
solar time 52
Sombrero Galaxy 90
SOUTHERN CROSS see CRUX
southern molecular cloud 78
space probes *137*, *141*, *143*, 144, 154
spectroscopic binaries 76, 83
Spica 50, 58, *58*, 60, *60*, 86, 88, 92
star trails *16*, *17*, *54*
stars: brightness 24–25, 66
  lifespan 30, 32–33
  to photograph *16*, 17
  spectra 26–27, *26*, 42
stellar wind 31
Stingray Nebula *21*
Straight Range 128
Straight Wall 130
Summer Triangle 58, 60, 96, 100, 102, 104
Sun 23, *25*, *37*, *54*, 117, **118–121**, 137
  chromosphere 118, 120, *120*, 121
  corona 118, 120, *120*, 121
  photosphere 118
  prominences 118, *119*, 120
  solar flares *119*
  solar wind 118
  sunquakes *119*
  sunspots 118, *118*
superclusters 45
supergiants 23, *25*, 33, 58, 78, 92, 94, 102

supernova remnants 102, *102*, 114
supernovae 25, *32*, 33, *33*, 72
Sword Handle see Double Cluster

T

Tarantula Nebula 38, *39*, 72
TAURUS 35, 51, 56, 62, 110, 112, **114**
  Alpha see Aldebaran
Taygete 114
telescopes 12–13, *12*, *13*
  giant 14–15
  radio 18, *18*
  space 19; see also Hubble Space Telescope
Tharsis Ridge 145
Theophilus crater 134
Thuban 68
Titan 150, *150*
Tombaugh, Clyde 149
Trapezium *51*, 78
TRIANGULUM **110–111**
Triangulum Galaxy 39
Trifid Nebula 28, *29*, 98, *98*
Triton 151, *151*
Tsiolkovsky crater *123*
TUCANA 62, **72**
Turtle Nebula 93
Tycho crater *126*, 130, *130*, 131

U

Uranus **148**, *148*
  moons *148*, 151
URSA MAJOR 34, 48, 50, 68, 70, **84**
  Alpha see Merak
  Beta see Dubhe
  Delta see Megrez
  Epsilon see Alioth
  Eta see Alkaid
  Gamma see Phad
  Zeta see Mizar
  see also BIG DIPPER
URSA MINOR **68**
  Alpha see Polaris

V

Valles Marineris 145
variable stars 25, 34, 66, 92
Vega 58, 60, 92, 96, 100
Veil Nebula 102, *102*
VELA 36, **83**
  stars within system 51, 60, 72, 83
Vendelinus crater 134
Venus 12, 122, 137, 139, 140, 142, **143**, *143*, 150
Very Large Array 18
Very Large Telescope 14–15, *15*
Vesta 154–155
VIRGO 86, **90**
  see also Spica
Virgo cluster 45, 88
Virgo supercluster 45
VULPECULA **101**

W

Walter crater 134
Whirlpool Galaxy 86, *87*
white dwarfs 32, *32*, 76, 80, 97, *101*, 108, 112
Wild Duck cluster 97, 100
Winking Demon see Algol
Winter Triangle 60
Wolf-Rayet stars 27, 83

Y

yellow dwarf 118

Z

zenith 52
zodiac see individual constellations, e.g. CANCER
Zubenelgenubi 89
Zubenelchemale 89

Acknowledgments

The author and publishers would like to thank Spacecharts Photo Library for picture research and for supplying many of the photographs in the book. They are also indebted to the following establishments for providing invaluable information and illustrative material: Anglo-Australian Observatory, Astronaut Memorial Planetarium and Observatory, European Southern Observatory, European Space Agency, Goddard Space Center, Jet Propulsion Laboratory, Kitt Peak National Observatory, London Planetarium (p.35), National Aeronautics and Space Administration, National Radio Astronomy Observatories, Palomar Observatory, Parkes Radio Observatory, Roque de los Muchachos Observatory, Royal Astronomical Society, Royal Greenwich Observatory, Space Telescope Science Institute, US Naval Observatory. Photographs on the following pages were taken by the author: 12, 13, 14b, 16, 18b and r, 120, 121, 152b.

While every effort has been made to trace and acknowledge all copyright holders, we would like to apologize should there have been any omissions. The publishers would also like to thank Pamela Ellis for supplying the index.